Advances in Middle Infrared Laser Crystals and Its Applications

Advances in Middle Infrared Laser Crystals and Its Applications

Editors

Xiaoming Duan
Renqin Dou
Linjun Li
Xiaotao Yang

MDPI • Basel • Beijing • Wuhan • Barcelona • Belgrade • Manchester • Tokyo • Cluj • Tianjin

Editors

Xiaoming Duan
Harbin Institute of
Technology
China

Renqin Dou
Chinese Academy of Sciences
China

Linjun Li
Harbin University of Science
and Technology
China

Xiaotao Yang
Harbin Engineering
University
China

Editorial Office
MDPI
St. Alban-Anlage 66
4052 Basel, Switzerland

This is a reprint of articles from the Special Issue published online in the open access journal *Crystals* (ISSN 2073-4352) (available at: https://www.mdpi.com/journal/crystals/special_issues/middle_infrared).

For citation purposes, cite each article independently as indicated on the article page online and as indicated below:

LastName, A.A.; LastName, B.B.; LastName, C.C. Article Title. *Journal Name* **Year**, *Volume Number*, Page Range.

ISBN 978-3-0365-4775-6 (Hbk)
ISBN 978-3-0365-4776-3 (PDF)

© 2022 by the authors. Articles in this book are Open Access and distributed under the Creative Commons Attribution (CC BY) license, which allows users to download, copy and build upon published articles, as long as the author and publisher are properly credited, which ensures maximum dissemination and a wider impact of our publications.

The book as a whole is distributed by MDPI under the terms and conditions of the Creative Commons license CC BY-NC-ND.

Contents

Xiaoming Duan, Renqin Dou, Linjun Li and Xiaotao Yang
Editorial for the Special Issue on "Advances in Middle Infrared Laser Crystals and Its Applications"
Reprinted from: *Crystals* **2022**, *12*, 643, doi:10.3390/cryst12050643 . . . . **1**

Yuhai Li, Cheng Zhang and Qinggang Ji
Efficient Ho:YVO$_4$ Laser Double-Pass-Pumped by a Wavelength-Stabilized Laser Diode
Reprinted from: *Crystals* **2022**, *12*, 320, doi:10.3390/cryst12030320 . . . . **5**

Nikolai Yudin, Andrei Khudoley, Mikhail Zinoviev, Sergey Podzvalov, Elena Slyunko, Elena Zhuravleva, Maxim Kulesh, Gennadij Gorodkin, Pavel Kumeysha and Oleg Antipov
The Influence of Angstrom-Scale Roughness on the Laser-Induced Damage Threshold of Single-Crystal ZnGeP$_2$
Reprinted from: *Crystals* **2022**, *12*, 83, doi:10.3390/cryst12010083 . . . . **11**

Nikolai Nikolayevich Yudin, Mikhail Zinoviev, Vladislav Gladkiy, Evgeny Moskvichev, Igor Kinyaevsky, Sergey Podzyvalov, Elena Slyunko, Elena Zhuravleva, Anastasia Pfaf, Nikolai Aleksandrovich Yudin and Maxim Kulesh
Influence of the Characteristics of Multilayer Interference Antireflection Coatings Based on Nb, Si, and Al Oxides on the Laser-Induced Damage Threshold of ZnGeP$_2$ Single Crystal
Reprinted from: *Crystals* **2021**, *11*, 1549, doi:10.3390/cryst11121549 . . . . **23**

Baizhong Li, Qiudi Chen, Peixiong Zhang, Ruifeng Tian, Lu Zhang, Qinglin Sai, Bin Wang, Mingyan Pan, Youchen Liu, Changtai Xia, Zhenqiang Chen and Hongji Qi
β-Ga$_2$O$_3$ Used as a Saturable Sbsorber to Realize Passively Q-Switched Laser Output
Reprinted from: *Crystals* **2021**, *11*, 1501, doi:10.3390/cryst11121501 . . . . **35**

Jiayu Liao, Qiudi Chen, Xiaochen Niu, Peixiong Zhang, Huiyu Tan, Fengkai Ma, Zhen Li, Siqi Zhu, Yin Hang, Qiguo Yang and Zhenqiang Chen
Energy Transfer and Cross-Relaxation Induced Efficient 2.78 μm Emission in Er^{3+}/Tm^{3+}: PbF$_2$ mid-Infrared Laser Crystal
Reprinted from: *Crystals* **2021**, *11*, 1024, doi:10.3390/cryst11091024 . . . . **43**

Lixue Wang, Xudong Sun, Congrui Geng, Zequn Zhang and Jixing Cai
A Study of the Characteristics of Plasma Generated by Infrared Pulse Laser-Induced Fused Silica
Reprinted from: *Crystals* **2021**, *11*, 1009, doi:10.3390/cryst11081009 . . . . **57**

Ilhwan Kim, Donghwa Lee and Kwang Jo Lee
Investigation of Mid-Infrared Broadband Second-Harmonic Generation in Non-Oxide Nonlinear Optic Crystals
Reprinted from: *Crystals* **2021**, *11*, 921, doi:10.3390/cryst11080921 . . . . **69**

Lihe Zheng, Jianbin Zhao, Yangxiao Wang, Weichao Chen, Fangfang Ruan, Hui Lin, Yanyan Xue, Jian Liu, Yang Liu, Ruiqin Yang, Haifeng Lu, Xiaodong Xu and Liangbi Su
Mid-IR Optical Property of Dy:CaF$_2$-SrF$_2$ Crystal Fabricated by Multicrucible Temperature Gradient Technology
Reprinted from: *Crystals* **2021**, *11*, 907, doi:10.3390/cryst11080907 . . . . **91**

Liang Chen, Di Wang, Guang-Yong Jin and Zhi Wei
Study on the Internal Mechanism of APD Photocurrent Characteristics Caused by the ms Pulsed Infrared Laser Irradiation
Reprinted from: *Crystals* **2021**, *11*, 884, doi:10.3390/cryst11080884 . . . . **101**

Liang Chen, Zhi Wei, Di Wang, Hong-Xu Liu and Guang-Yong Jin
Temperature Rise Characteristics of Silicon Avalanche Photodiodes in Different External Capacitance Circuits Irradiated by Infrared Millisecond Pulse Laser
Reprinted from: *Crystals* **2021**, *11*, 866, doi:10.3390/cryst11080866 . . . . **107**

Dayong Zhang, Li Cheng and Zuochun Shen
Formation Laws of Direction of Fano Line-Shape in a Ring MIM Plasmonic Waveguide Side-Coupled with a Rectangular Resonator and Nano-Sensing Analysis of Multiple Fano Resonances
Reprinted from: *Crystals* **2021**, *11*, 819, doi:10.3390/cryst11070819 . . . . **113**

Yanqiu Du, Tongyu Dai, Hui Sun, Hui Kang, Hongyang Xia, Jiaqi Tian, Xia Chen and Baoquan Yao
Experimental Investigation of Double-End Pumped Tm, Ho: $GdVO_4$ Laser at Cryogenic Temperature
Reprinted from: *Crystals* **2021**, *11*, 798, doi:10.3390/cryst11070798 . . . . **125**

Youbao Ni, Qianqian Hu, Haixin Wu, Weimin Han, Xuezhou Yu and Mingsheng Mao
The Investigation on Mid-Far Infrared Nonlinear Crystal $AgGaGe_5Se_{12}$ (AGGSe)
Reprinted from: *Crystals* **2021**, *11*, 661, doi:10.3390/cryst11060661 . . . . **133**

Chuanpeng Qian, Ting Yu, Jing Liu, Yuyao Jiang, Sijie Wang, Xiangchun Shi, Xisheng Ye and Weibiao Chen
A High-Energy, Narrow-Pulse-Width, Long-Wave Infrared Laser Based on ZGP Crystal
Reprinted from: *Crystals* **2021**, *11*, 656, doi:10.3390/cryst11060656 . . . . **139**

Chao Niu, Yan Jiang, Ya Wen, Lu Zhao, Xinyu Chen, Chunting Wu and Tongyu Dai
High-Efficiency Ho:YAP Pulse Laser Pumped at 1989 nm
Reprinted from: *Crystals* **2021**, *11*, 595, doi:10.3390/cryst11060595 . . . . **147**

Yao Ma, Chao Xin, Wei Zhang and Guangyong Jin
Experimental Study of Plasma Plume Analysis of Long Pulse Laser Irradiates CFRP and GFRP Composite Materials
Reprinted from: *Crystals* **2021**, *11*, 545, doi:10.3390/cryst11050545 . . . . **157**

Editorial

Editorial for the Special Issue on "Advances in Middle Infrared Laser Crystals and Its Applications"

Xiaoming Duan [1,*], Renqin Dou [2,*], Linjun Li [3,*] and Xiaotao Yang [4,*]

1. National Key Laboratory of Tunable Laser Technology, Harbin Institute of Technology, Harbin 150001, China
2. The Key Laboratory of Photonic Devices and Materials, Anhui Institute of Optics and Fine Mechanics, Chinese Academy of Sciences, Hefei 230001, China
3. Heilongjiang Province Key Laboratory of Laser Spectroscopy Technology and Application, Harbin University of Science and Technology, Harbin 150001, China
4. College of Power and Energy Engineering, Harbin Engineering University, Harbin 150001, China
* Correspondence: xmduan@hit.edu.cn (X.D.); drq0564@aiofm.ac.cn (R.D.); llj7897@126.com (L.L.); yangxiaotao@hrbeu.edu.cn (X.Y.)

Citation: Duan, X.; Dou, R.; Li, L.; Yang, X. Editorial for the Special Issue on "Advances in Middle Infrared Laser Crystals and Its Applications". *Crystals* 2022, *12*, 643. https://doi.org/10.3390/cryst12050643

Received: 25 April 2022
Accepted: 26 April 2022
Published: 30 April 2022

Publisher's Note: MDPI stays neutral with regard to jurisdictional claims in published maps and institutional affiliations.

Copyright: © 2022 by the authors. Licensee MDPI, Basel, Switzerland. This article is an open access article distributed under the terms and conditions of the Creative Commons Attribution (CC BY) license (https://creativecommons.org/licenses/by/4.0/).

In the past two decades, there has been a growing interest in middle infrared (mid-IR) laser crystals and its application to achieve mid-IR laser radiations, which has been benefited by the development of novel mid-infrared crystals and the improving quality of traditional mid-IR crystals. The recent progress in crystal growth, theoretical modelling, and generation of mid-IR laser radiations has offered a new perspective for design and growth of mid-IR crystals. In this Special Issue of *Crystals*, we gathered sixteen peer-reviewed papers that shed light on recent advances in the field of mid-IR laser crystals and their applications.

Li et al. reported the lasing performance of a Ho:YVO$_4$ crystal under laser diode (LD) pumping conditions [1]. With double-pass-pumped architecture, up to 8.7 W continuous-wave (CW) output power at 2052.4 nm was achieved. In addition, good beam quality factors in horizontal and vertical directions were also obtained. Yudin et al. explored the influence of angstrom-scale roughness on the laser-induced damage threshold of ZnGeP$_2$ crystal [2]. In the polishing process, the magnetorheological processing technology was used. Samples of the ZnGeP$_2$ with an Angstrom level of surface roughness were achieved. The laser-induced damage threshold value at the indicated orders of magnitude of the surface roughness parameters was determined by the number of point depressions. The influence on the characteristics of multilayer Interference antireflection coatings was also investigated, based on Nb, Si, and Al oxides on the laser-induced damage threshold of ZnGeP$_2$ crystal [3]. Under the excitation of a 2.09-μm Ho:YAG laser, the effect of the defect structure and the parameters of antireflection interference coatings based on alternating layers of Nb$_2$O$_5$/Al$_2$O$_3$ and Nb$_2$O$_5$/SiO$_2$ layers on the laser-induced damage threshold of ZnGeP$_2$ crystals were determined. Experimental results indicate that the silicon conglomerates in an interference antireflection coating is beneficial to decrease the laser-induced damage threshold of the ZnGeP$_2$ crystal. Li et al. prepared the β-Ga$_2$O$_3$ saturable absorber by the optical floating zone method [4], which was used in subsequent work, the 1.08-μm Nd:GYAP laser produces 606.54-ns pulse width at repetition rate of 344.06 kHz. Experimental results present that the β-Ga$_2$O$_3$ crystal has great potential for the development of the 1-μm pulsed laser. Liao et al. explored the energy transfer and cross-relaxation induced 2.78 μm emission in Er^{3+}/Tm^{3+}: PbF$_2$ crystal [5], which was grown by the Bridgman method. The spectroscopic properties, energy transfer mechanism, and first-principles calculations of as-grown crystals were investigated in detail. This work provides an avenue to design mid-IR laser materials with good performance. Wang et al. investigated the characteristics of plasma generated by infrared pulse laser-induced fused silica [6]. Based on the theory of fluid mechanics and gas dynamics, a two-dimensional axisymmetric gas dynamic model was established to simulate the plasma generation process of fused silica induced by a millisecond pulse laser. The maximum expansion velocity of the laser-induced plasma was calculated. The experimental results verify the correctness of theoretical calculations. Kim et al. theoretically and numerically investigated the broadband second-harmonic properties of non-oxide mid-IR nonlinear crystals [7]. Chalcopyrite semiconductors, defect chalcopyrite and orthorhombic ternary chalcogenides have also been demonstrated. Beam propagation directions, spectral positions of resonance, effective nonlinearities, spatial walk-offs between interacting beams and spectral bandwidths were analyzed. Zheng et al. investigated the mid-IR optical property of Dy:CaF$_2$-SrF$_2$ crystal [8], which was fabricated by a multicrucible temperature gradient method. Sellmeier dispersion formula, absorption characteristics and fluorescence properties were demonstrated. The results indicate that the Dy:CaF$_2$-SrF$_2$ crystal is a promising candidate for compact mid-IR lasers. Chen et al. show the internal mechanism of APD photocurrent characteristics under ms-level pulsed infrared laser irradiation [9]. The sampling current characteristics of the external circuit and the internal mechanism of the current generation in APD were studied. Chen et al. also investigated the temperature-rise characteristics of silicon avalanche photodiodes in different external capacitance circuits irradiated by the same laser pulse [10]. Zhang et al. explored the formation laws of direction of Fano line-shape in a ring MIM plasmonic waveguide side-coupled with a rectangular resonator and nano-sensing analysis of multiple fano resonances [11]. Du et al. presented an experimental investigation of double-end pumped Tm, Ho:GdVO$_4$ laser [12]. With operating temperature of 77 K, the output characteristics of Tm, Ho:GdVO$_4$ laser were studied, where the continuous-wave and Q-switching performance were also demonstrated. Ni et al. investigated the mid-far properties of nonlinear crystal AGGSe [13]. The large AGGSe single crystal of 35 mm diameter and 80 mm length was obtained by the seed-aided Bridgman method. X-ray diffraction, rocking curve and transmission spectrum was used to characterize the crystalline quality. In addition, the 8 μm frequency doubling was also presented. Qian et al. reported a high energy, narrow pulse width, long-wave infrared ZGP optical parametric oscillator [14]. The maximum average output powers of 3.15 W at 8.2 μm and 11.4 W at 2.8 μm were achieved. The minimum pulse width was approximately 8.1 ns. Niu et al. reported an efficient pulsed Ho:YAP laser pumped at 1989 nm [15]. When the absorbed power was 30 W, an average power of 18.02 W with the pulse width of 104.2 ns acousto-optic (AO) Q-switched Ho:YAP laser was obtained at a repetition frequency of 10 kHz. Ma et al. experimentally investigated the plasma plume analysis of long pulse laser irradiates CFRP and GFRP composite materials [16]. The results show that GFRP is more vulnerable to breakdown than CFRP under the same conditions.

As shown in this Special Issue of *Crystals*, the study of mid-IR crystals and its applications continues to grow and expand as we, as a community, strive to acquire further understanding of the underlying potential of these crystals. The goal is to bring these and other new concepts closer to application for mid-IR crystals and beyond.

Funding: This research received no external funding.

Conflicts of Interest: The authors declare no conflict of interest.

References

1. Li, Y.; Zhang, C.; Ji, Q. Efficient Ho:YVO$_4$ Laser Double-Pass-Pumped by a Wavelength-Stabilized Laser Diode. *Crystals* **2022**, *12*, 320. [CrossRef]
2. Yudin, N.; Khudoley, A.; Zinoviev, M.; Podzvalov, S.; Slyunko, E.; Zhuravleva, E.; Kulesh, M.; Gorodkin, G.; Kumeysha, P.; Antipov, O. The Influence of Angstrom-Scale Roughness on the Laser-Induced Damage Threshold of Single-Crystal ZnGeP$_2$. *Crystals* **2022**, *12*, 83. [CrossRef]
3. Yudin, N.N.; Zinoviev, M.; Gladkiy, V.; Moskvichev, E.; Kinyaevsky, I.; Podzyvalov, S.; Slyunko, E.; Zhuravleva, E.; Pfaf, A.; Yudin, N.A.; et al. Influence of the Characteristics of Multilayer Interference Antireflection Coatings Based on Nb, Si, and Al Oxides on the Laser-Induced Damage Threshold of ZnGeP$_2$ Single Crystal. *Crystals* **2021**, *11*, 1549. [CrossRef]
4. Li, B.; Chen, Q.; Zhang, P.; Tian, R.; Zhang, L.; Sai, Q.; Wang, B.; Pan, M.; Liu, Y.; Xia, C.; et al. β-Ga$_2$O$_3$ Used as a Saturable Sbsorber to Realize Passively Q-Switched Laser Output. *Crystals* **2021**, *11*, 1501. [CrossRef]

5. Liao, J.; Chen, Q.; Niu, X.; Zhang, P.; Tan, H.; Ma, F.; Li, Z.; Zhu, S.; Hang, Y.; Yang, Q.; et al. Energy Transfer and Cross-Relaxation Induced Efficient 2.78 μm Emission in Er^{3+}/Tm^{3+}: PbF_2 mid-Infrared Laser Crystal. *Crystals* **2021**, *11*, 1024. [CrossRef]
6. Wang, L.; Sun, X.; Geng, C.; Zhang, Z.; Cai, J. A Study of the Characteristics of Plasma Generated by Infrared Pulse Laser-Induced Fused Silica. *Crystals* **2021**, *11*, 1009. [CrossRef]
7. Kim, I.; Lee, D.; Lee, K.J. Investigation of Mid-Infrared Broadband Second-Harmonic Generation in Non-Oxide Nonlinear Optic Crystals. *Crystals* **2021**, *11*, 921. [CrossRef]
8. Zheng, L.; Zhao, J.; Wang, Y.; Chen, W.; Ruan, F.; Lin, H.; Xue, Y.; Liu, J.; Liu, Y.; Yang, R.; et al. Mid-IR Optical Property of Dy:CaF_2-SrF_2 Crystal Fabricated by Multicrucible Temperature Gradient Technology. *Crystals* **2021**, *11*, 907. [CrossRef]
9. Chen, L.; Wang, D.; Jin, G.-Y.; Wei, Z. Study on the Internal Mechanism of APD Photocurrent Characteristics Caused by the ms Pulsed Infrared Laser Irradiation. *Crystals* **2021**, *11*, 884. [CrossRef] .
10. Chen, L.; Wei, Z.; Wang, D.; Liu, H.-X.; Jin, G.-Y. Temperature Rise Characteristics of Silicon Avalanche Photodiodes in Different External Capacitance Circuits Irradiated by Infrared Millisecond Pulse Laser. *Crystals* **2021**, *11*, 866. [CrossRef]
11. Zhang, D.; Cheng, L.; Shen, Z. Formation Laws of Direction of Fano Line-Shape in a Ring MIM Plasmonic Waveguide Side-Coupled with a Rectangular Resonator and Nano-Sensing Analysis of Multiple Fano Resonances. *Crystals* **2021**, *11*, 819. [CrossRef]
12. Du, Y.; Dai, T.; Sun, H.; Kang, H.; Xia, H.; Tian, J.; Chen, X.; Yao, B. Experimental Investigation of Double-End Pumped Tm, Ho: GdVO4 Laser at Cryogenic Temperature. *Crystals* **2021**, *11*, 798. [CrossRef]
13. Ni, Y.; Hu, Q.; Wu, H.; Han, W.; Yu, X.; Mao, M. The Investigation on Mid-Far Infrared Nonlinear Crystal AgGaGe5Se12 (AGGSe). *Crystals* **2021**, *11*, 661. [CrossRef]
14. Qian, C.; Yu, T.; Liu, J.; Jiang, Y.; Wang, S.; Shi, X.; Ye, X.; Chen, W. A High-Energy, Narrow-Pulse-Width, Long-Wave Infrared Laser Based on ZGP Crystal. *Crystals* **2021**, *11*, 656. [CrossRef]
15. Niu, C.; Jiang, Y.; Wen, Y.; Zhao, L.; Chen, X.; Wu, C.; Dai, T. High-Efficiency Ho:YAP Pulse Laser Pumped at 1989 nm. *Crystals* **2021**, *11*, 595. [CrossRef]
16. Ma, Y.; Xin, C.; Zhang, W.; Jin, G. Experimental Study of Plasma Plume Analysis of Long Pulse Laser Irradiates CFRP and GFRP Composite Materials. *Crystals* **2021**, *11*, 545. [CrossRef]

Article

Efficient Ho:YVO$_4$ Laser Double-Pass-Pumped by a Wavelength-Stabilized Laser Diode

Yuhai Li [1,2,*], Cheng Zhang [3] and Qinggang Ji [2]

1. School of Precision Instrument and Opto-Electronics Engineering, Tianjin University, Tianjin 300072, China
2. Science and Technology on Electro-Optical Information Security Control Laboratory, Academy of Opto-Electronics of China Electronics Technology Group Corporation, Tianjin 300308, China; jiqinggang53@163.com
3. Shenyang Aircraft Design & Research Institute, Shenyang 110035, China; zhangcheng811@126.com
* Correspondence: aoe-cetc@cetc.com.cn

Abstract: A Ho:YVO$_4$ laser double-pass-pumped with a 1.91 μm wavelength-stabilized laser diode is presented in this paper. The maximum output power was up to 8.7 W, with a wavelength of 2052.4 nm and a slope efficiency of 37.4%. The M^2 factors in the x and y directions were 1.8 and 1.6 at the maximum output power, respectively.

Keywords: laser; Ho:YVO$_4$; double-pass-pumping; laser diode

Citation: Li, Y.; Zhang, C.; Ji, Q. Efficient Ho:YVO$_4$ Laser Double-Pass-Pumped by a Wavelength-Stabilized Laser Diode. *Crystals* **2022**, *12*, 320. https://doi.org/10.3390/cryst12030320

Academic Editors: Carlo Vicario, Xiaoming Duan, Renqin Dou, Linjun Li and Xiaotao Yang

Received: 29 October 2021
Accepted: 11 February 2022
Published: 25 February 2022

Publisher's Note: MDPI stays neutral with regard to jurisdictional claims in published maps and institutional affiliations.

Copyright: © 2022 by the authors. Licensee MDPI, Basel, Switzerland. This article is an open access article distributed under the terms and conditions of the Creative Commons Attribution (CC BY) license (https://creativecommons.org/licenses/by/4.0/).

1. Introduction

Today, 2 μm solid-state lasers have many applications in optical measurement, wind finding lidars, atmospheric monitoring, space communication, and medical treatment. Diode-pumped Tm, Ho co-doped gain-mediums, which take advantage of the two-for-one process pumped by ~792 nm, have been used extensively to yield 2 μm lasers. However, Tm, Ho co-doped lasers with quasi-three-level rely on energy-transfer processes and exhibit losses in radiation and irradiation, which lead to large heat loading of the gain-mediums in room-temperature operation [1]. In order to yield more powerful and brighter lasers, liquid nitrogen temperature is always needed [2]. In contrast, singly Ho-doped lasers by direct-resonant-pumping [3–10] to obtain 2 μm lasers have the advantage of high conversion efficiency and less thermal loading because of the weaker quantum defect between the laser and pump; thus, operation at room temperature is possible.

The spectral characteristics of the Ho:YVO$_4$ crystal around 2 μm at room temperature have been previously reported [11]. The long fluorescence decay lifetime and the large emission cross section make the Ho:YVO$_4$ crystal an outstanding host for 2 μm lasers. Moreover, the large absorption cross section corresponding to 1.94 μm makes the Ho:YVO$_4$ crystal efficiently and resonantly pumped by 1.9 μm Tm-fiber and Tm-bulk lasers, and the continuous wave and Q-switching operations of the Ho:YVO$_4$ laser have been also demonstrated under the pumping of FBG-locked Tm-fiber and Tm:YAP bulk lasers [12–14]. A cryogenically cooled Tm, Ho:GdVO$_4$ laser with an output power of 7.4 W was reported by Du et al., which required the assistance of sophisticated cryogenic cooling systems at a temperature of 77 K [15]. An efficient room-temperature Q-switched Ho:YVO$_4$ laser pumped by a 1940 nm Tm-fiber laser was reported by Ding et al., which generated an average output power of 11.4 W [16]. The beam quality of the Tm-fiber pumped Ho:YVO$_4$ laser was better than the laser in this present work because fiber lasers has better optical output characteristics than semiconductor lasers.

In this paper, we report the first (as far as we know) Ho:YVO$_4$ laser double-pass-pumped by a 1.91 μm wavelength-stabilized laser diode (LD). At a total absorbed pump power of 30.2 W, the maximum output power was up to 8.7 W, with a central wavelength of 2052.4 nm and slope efficiency of 37.4% with the absorbed pump power. The beam quality (M^2) factors were 1.8 and 1.6 in the x and y directions, respectively.

2. Experimental Setup

Figure 1 shows the absorption cross sections of the Ho:YVO$_4$ crystal and the radiation spectrum of the LD. The strongest absorption peak is located at 1.94 μm, as in many previous reports. However, the strong absorption of the Ho:YVO$_4$ crystal led to serious thermal loading. Thus, the slightly weaker absorption of 1.91 μm was selected in this experiment, which could reduce the absorption and improve the thermal distribution uniformity.

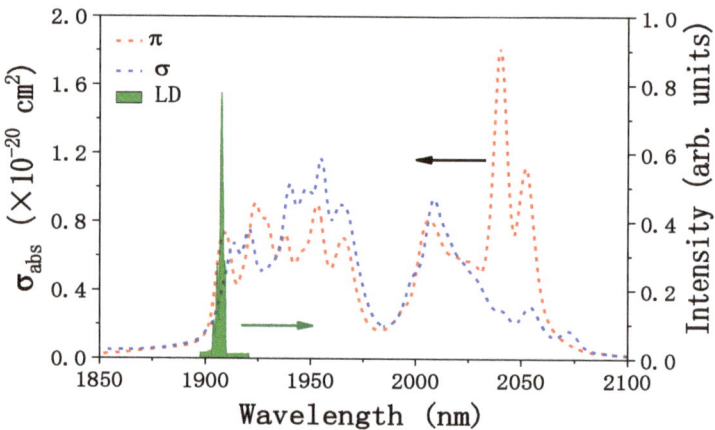

Figure 1. Absorption cross sections of the Ho:YVO$_4$ crystal and the radiation spectrum of the LD.

Figure 2 shows the experimental scheme of the LD double-pass-pumped Ho:YVO$_4$ laser. A wavelength-stabilized and fiber-coupled LD (QPC Corp., Los Angeles, USA) was employed as the pump, the core diameter and *NA* of which were 600 μm and 0.22, respectively. The central wavelength of the LD was 1.91 μm, with a linewidth of about 2 nm (FWHM) at the maximum output power of 40 W. The Ho:YVO$_4$ crystal was *a*-cut, the dimension of which was 3 × 3 mm^2 in cross section and 30 mm in length, and the doping concentration of the Ho^{3+} was 0.5 at.%. The crystal, both end-faces of which were anti-reflection coated at 1.9~2.1 μm, was wrapped in a heat sink made of copper and controlled at 15 °C using a thermoelectric cooler (Tecooler technology Co., Ltd., Shenzhen, China). The pump beam was reshaped with the focal lens of F1 and F2 and shot into the center of the crystal (with a radius of about 0.3 mm). The single-pass absorption of the crystal corresponding to the 1.91 μm pump was measured to be 51% when the cavity was absent. Although the overall cost of the above scheme is a little high, we can still accept that. We believe that with the improvement of semiconductor technology, the price of semiconductor lasers will become lower and lower.

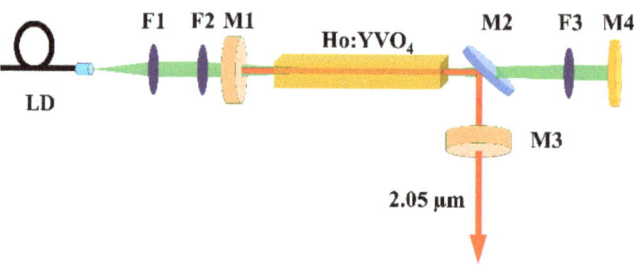

Figure 2. Experimental scheme of the Ho:YVO$_4$ laser.

The cavity (with a physical length of 50 mm) was L-shaped and consisted of an input mirror M1, a 45° reflectance mirror M2, and an output mirror M3, which were coated with high transmittance at 1.91 µm ($T > 99.97\%$) and high reflectance at 2.05 µm ($R > 99.98\%$), high transmittance at 1.91 µm ($T > 99.97\%$) and high reflectance at 2.05 µm ($T > 99.98\%$) with an angle of 45°, and partial reflectance at 2.05 µm, respectively. M1 and M2 were flat mirrors, whilst the plano-concave M3 had a curvature radius of 500 mm. A flat mirror M4 with high reflectance at 1.91 µm and a focal lens F3 (focal length of 30 mm) were employed to reflect the pump back to the crystal. In this way, the total pump absorption was increased to about 76%.

3. Experimental Results

Figure 3 shows that the output power depends on the absorbed pump power of the Ho:YVO$_4$ laser. With an output mirror transmittance of 10%, the maximum output power was 5.6 W with respect to the absorbed pump power of 30.2 W, corresponding to a slope efficiency of 23.6%. When the output mirror transmittance was 30%, the maximum output power and slope efficiency increased to 7.9 W and 34%, respectively. The maximum output power and slope efficiency reached the optimum values of 8.7 W and 37.4%, respectively, in the case of an output mirror transmittance of 50%. Although using diode pumping instead of pumping with a 1.04 µm laser based on a Tm-doped fiber or a Tm-doped crystal and via a double-pass pumping scheme is novel here, the efficiencies of the above scheme are lower than expected. In future work, we will use a 1940 nm diode laser instead of the present 1910 nm diode laser, increasing the single-pass absorption of the laser crystal and reducing the non-radiative loss, quantum defects, and thermal effects, which are all beneficial for improving the conversion efficiency.

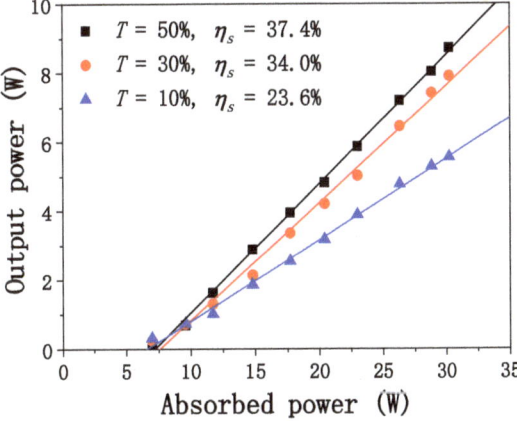

Figure 3. Output power dependence on the absorbed pump power of the Ho:YVO$_4$ laser.

The output spectrums of the Ho:YVO$_4$ laser at different transmittances of the output mirrors (shown in the Figure 4) were measured with the spectrum analyzer Bristol 721A (with a resolution of ±0.2 ppm). The central wavelength was 2066.1 nm at an output mirror transmittance of 10%. The laser emission line width (FWHM) was less than 1 nm. With the other two transmittances, the central wavelengths were blue-shifted to 2052.4 nm, which can be attributed to the low resonator loss. With the three transmittances of the output mirrors, there were no other emission peaks observed in the experiment.

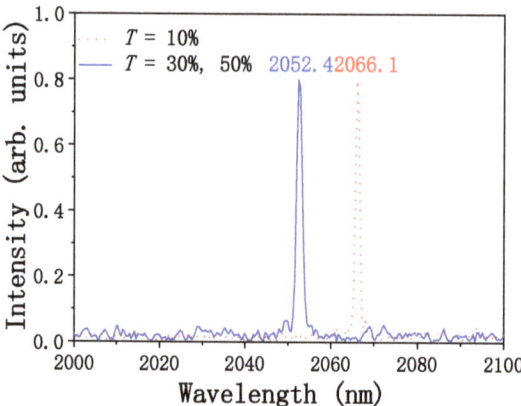

Figure 4. Output spectrum of the Ho:YVO$_4$ laser.

We measured the beam quality factor M^2 at the full output power of 8.7 W, taking advantage of the knife-edge method. Figure 5 shows the that laser beam radii depend on the location relative to the focal lens of 150 mm, which was used for leading out the waist of the oscillating beam in the cavity. The distance of the focal lens from the output coupler was about 100 mm. Using Gaussian fitting, the M^2 factors were calculated to be 1.8 and 1.6 in the x and y directions, respectively, which were better than those of the previous work, for example, Ref. [12]. Under the above experimental conditions, the stability of the laser resonator is judged by the ABCD matrix method [17] (the thermal focal length of the crystal was 500 mm at low pump power and 230 mm at high pump power).

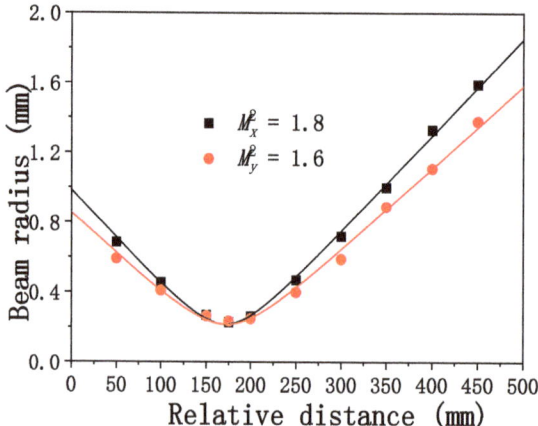

Figure 5. Beam quality of the Ho:YVO$_4$ laser.

4. Conclusions

Using a 1.91 μm wavelength-stabilized LD as the double-pass pump source, we demonstrated a continuous-wave Ho:YVO$_4$ laser. At an absorbed pump power of 30.2 W, the maximum output power was 8.7 W at 2052.4 nm, corresponding to a slope efficiency of 37.4%. The M^2 factors in the x and y directions were 1.8 and 1.6, respectively. The results imply that the LD double-pass-pumped Ho:YVO$_4$ laser is an efficient way to generate the 2 μm laser. In future work, we will use a 1940 nm LD instead of the present 1910 nm LD, increasing the single-pass absorption of the laser crystal and reducing the non-radiative loss, quantum defects, and thermal effects, which are all beneficial for improving the conversion efficiency.

Author Contributions: Data curation, Investigation, Methodology, Writing—original draft, Y.L.; Formal analysis, Resources, C.Z.; Validation, Writing—review & editing, Q.J. All authors have read and agreed to the published version of the manuscript.

Funding: This research received no external funding.

Institutional Review Board Statement: Not applicable.

Informed Consent Statement: Not applicable.

Data Availability Statement: Not applicable.

Conflicts of Interest: The authors declare no conflict of interest.

References

1. Rustad, G.; Stenersen, K. Modeling of laser-pumped Tm and Ho lasers accounting for upconversion and ground-state depletion. *IEEE J. Quantum Electron.* **1996**, *32*, 1645–1655. [CrossRef]
2. Li, L.J.; Bai, Y.F.; Duan, X.M.; Qin, J.P.; Wang, J.; He, Z.L.; Zhou, S.; Zhang, Z.G. A continuous-wave b-cut Tm, Ho:YAlO$_3$ laser with a 15 W output pumped by two laser diodes. *Laser Phys. Lett.* **2013**, *10*, 035802. [CrossRef]
3. Duan, X.M.; Yao, B.Q.; Song, C.W.; Gao, J.; Wang, Y.Z. Room temperature efficient continuous wave and Q-switched Ho:YAG laser double-pass pumped by a diode-pumped Tm:YLF laser. *Laser Phys. Lett.* **2008**, *5*, 800–803. [CrossRef]
4. Duan, X.M.; Yao, B.Q.; Li, G.; Wang, T.H.; Yang, X.T.; Wang, Y.Z.; Zhao, G.J.; Dong, Q. High efficient continuous wave operation of a Ho:YAP laser at room temperature. *Laser Phys. Lett.* **2009**, *6*, 279–281. [CrossRef]
5. Duan, X.M.; Yao, B.Q.; Li, G.; Ju, Y.L.; Wang, Y.Z.; Zhao, G.J. High efficient actively Q-switched Ho:LuAG laser. *Opt. Express* **2009**, *17*, 21691–21697. [CrossRef] [PubMed]
6. Kohei, M.; Shoken, I.; Makoto, A.; Hironori, I.; Ryohei, O.; Hirotake, F.; Takayoshi, I.; Atsushi, S. 2 μm Doppler wind lidar with a Tm:fiber-laser-pumped Ho:YLF laser. *Opt. Lett.* **2018**, *43*, 202–205.
7. Němec, M.; Šulc, J.; Jelínek, M.; Kubeček, V.; Jelínková, H.; Doroshenko, M.E.; Alimov, O.; Konyushkin, V.A.; Nakladov, A.N.; Osiko, V.V. Thulium fiber pumped tunable Ho:CaF$_2$ laser. *Opt. Lett.* **2017**, *42*, 1852–1855. [CrossRef] [PubMed]
8. Duan, X.M.; Shen, Y.J.; Gao, J.; Zhu, H.B.; Qian, C.P.; Su, L.B.; Zheng, L.H.; Li, L.J.; Yao, B.Q.; Dai, T.Y. Active Q-switching operation of slab Ho:SYSO laser wing-pumped by fiber coupled laser diodes. *Opt. Express* **2019**, *27*, 11455–11461. [CrossRef] [PubMed]
9. Kifle, E.; Loiko, P.; Romero, C.; Rodríguez, J.; Ródenas, A.; Zakharov, V.; Veniaminov, A.; Aguiló, M.; Díaz, F.; Griebner, U.; et al. Femtosecond-laser-written Ho:KGd(WO$_4$)$_2$ waveguide laser at 2.1 μm. *Opt. Lett.* **2019**, *44*, 1738–1741. [CrossRef] [PubMed]
10. Duan, X.M.; Wu, J.Z.; Dou, R.Q.; Zhang, Q.L.; Dai, T.Y.; Yang, X.T. High-power actively Q-switched Ho-doped gadolinium tantalate laser. *Opt. Express* **2021**, *29*, 12471–12477. [CrossRef] [PubMed]
11. Gołvab, S.; Solarz, P.; Dominiak-Dzik, G.; Lukasiewicz, T.; Świrkowicz, M.; Ryba-Romanowski, W. Spectroscopy of YVO$_4$:Ho^{3+} crystals. *Appl. Phys. B Lasers Opt.* **2002**, *74*, 237–241. [CrossRef]
12. Li, G.; Yao, B.Q.; Meng, P.B.; Ju, Y.L.; Wang, Y.Z. High-efficiency resonantly pumped room temperature Ho:YVO$_4$ laser. *Opt. Lett.* **2011**, *36*, 2934–2936. [CrossRef] [PubMed]
13. Han, L.; Yao, B.Q.; Duan, X.M.; Li, S.; Dai, T.Y.; Ju, Y.L.; Wang, Y.Z. Experimental study of continuous-wave and Q-switched laser performances of Ho:YVO$_4$ crystal. *Chin. Opt. Lett.* **2014**, *12*, 081401. [CrossRef]
14. Dai, T.Y.; Ding, Y.; Ju, Y.L.; Yao, B.Q.; Li, Y.Y.; Wang, Y.Z. High repetition frequency passively Q-switched Ho:YVO$_4$ laser. *Infrared Phys. Technol.* **2015**, *72*, 254–257. [CrossRef]
15. Du, Y.Q.; Dai, T.Y.; Sun, H.; Kang, H.; Xia, H.Y.; Tian, J.Q.; Chen, X.; Yao, B.Q. Experimental Investigation of Double-End Pumped Tm, Ho:GdVO$_4$ Laser at Cryogenic Temperature. *Crystals* **2021**, *11*, 798. [CrossRef]
16. Ding, Y.; Yao, B.Q.; Ju, Y.L.; Li, Y.Y.; Duan, X.M.; He, W.J. High power Q-switched Ho: YVO$_4$ laser resonantly pumped by a Tm-fiber-laser. *Laser Phys.* **2015**, *25*, 015002. [CrossRef]
17. Brandus, C.A.; Dascalu, T. Cavity design peculiarities and influence of SESAM characteristics on output performances of a Nd:YVO$_4$ mode locked laser oscillator. *Opt. Laser Technol.* **2019**, *111*, 452–458. [CrossRef]

Article

The Influence of Angstrom-Scale Roughness on the Laser-Induced Damage Threshold of Single-Crystal ZnGeP$_2$

Nikolai Yudin [1,*], Andrei Khudoley [2], Mikhail Zinoviev [1], Sergey Podzvalov [1], Elena Slyunko [1], Elena Zhuravleva [1], Maxim Kulesh [1], Gennadij Gorodkin [2], Pavel Kumeysha [2] and Oleg Antipov [3,4]

[1] Radiophysics Department, National Research Tomsk State University, 634050 Tomsk, Russia; muxa9229@gmail.com (M.Z.); cginen@yandex.ru (S.P.); elenohka266@mail.ru (E.S.); lenazhura@mail.ru (E.Z.); kyleschMM2000@yandex.ru (M.K.)
[2] A.V. Luikov Heat and Mass Transfer Institute NASB, 220072 Minsk, Belarus; khudoley@hmti.ac.by (A.K.); gorodkin@hmti.ac.by (G.G.); pavel.k@hmti.ac.by (P.K.)
[3] Institute of Applied Physics of the Russian Academy of Sciences, 603950 Nizhny Novgorod, Russia; antipov@ipfran.ru
[4] Higher School of General and Applied Physics, Nizhny Novgorod State University, 603950 Nizhny Novgorod, Russia
* Correspondence: rach3@yandex.ru; Tel.: +7-996-938-71-32

Abstract: Magnetorheological processing was applied to polish the working surfaces of single-crystal ZnGeP$_2$, in which a non-aqueous liquid with the magnetic particles of carbonyl iron with the addition of nanodiamonds was used. Samples of a single-crystal ZnGeP$_2$ with an Angstrom level of surface roughness were received. The use of magnetorheological polish allowed the more accurate characterization of the possible structural defects that emerged on the surface of a single crystal and had a size of ~0.5–1.5 μm. The laser-induced damage threshold (LIDT) value at the indicated orders of magnitude of the surface roughness parameters was determined not by the quality of polishing, but by the number of point depressions caused by the physical limitations of the structural configuration of the crystal volume. These results are in good agreement with the assumption made about a significant effect of the concentration of dislocations in a ZnGeP$_2$ crystal on LIDT.

Keywords: laser-induced damage threshold; ZnGeP$_2$; magnetorheological polish

1. Introduction

Repetitively pulsed coherent powerful radiation sources in the mid-IR range have a wide variety of applications in many areas, such as material processing (glass, ceramics, or semiconductors) [1,2] and medicine, including disease diagnosis using gas analysis and the resonance ablation of biological tissues [3]. Coherent radiation sources capable of generating powerful pulsed radiation in the wavelength range of 3.5–5 μm are relevant when creating lidar systems based on the differential absorption method for the control of greenhouse gas emissions (as the most intensive absorption lines of greenhouse gases are in this spectral range) [4–6]. Among the most effective solid sources of coherent radiation in the mid-IR range are optical parametric oscillators (OPO).

Currently, the most powerful OPOs in the wavelength range of 3.5–5 μm are developed based on nonlinear-optical ZnGeP$_2$ (ZGP) crystals [7]. The OPO data can generate radiation with an average power of up to 160 W, or pulse energy up to 200 mJ with a pulse duration of 20–60 ns and a repetition rate of up to 100 kHz [8–10]. However, long-term work without the failure of powerful OPOs based on ZGP is limited to the laser-induced damage threshold (LIDT) of the surface of the given material. In this regard, the potential for practical use of the OPO data of the mid-IR range is associated with the need to improve the methods for the processing of the working surfaces of crystals to increase their LIDT. The problem of ZGP optical breakdown by laser radiation at wavelengths from 1.064 μm to 10 µm is the subject of several published articles [11–18]. These articles revealed a significant difference in the magnitude of the LIDT of the ZGP crystal at the wavelengths of 1.064 µm and 2.1 µm [11]. The dynamic visualization of the breakdown process with laser radiation at a wavelength of 2.1 µm in the volume of ZGP showed that an avalanche-shaped temperature increase occurs within the nonlinear-optical element [12]. An increase in the ZGP breakdown threshold with a decrease in the duration of the pump radiation pulses is "indicated in favor of the thermal nature of the breakdown for nanosecond pulses due to abnormal infrared absorption." In [14], it was shown that the cooling of a crystal to a temperature of −60 °C leads to an increase in the LIDT by 1.5–3 times, up to 9 J/cm^2 at the wavelength of the acting laser radiation of 2.091 µm and the frequency of 10 kHz pulses.

In [15], it was reported that the LIDT of ZGP elements at a wavelength of 9.55 µm was determined by the intensity of the acting beam of 142 MW/cm^2, a pulse duration of 85 ns and a repetition rate of 1 Hz, which is ~9.5 J/cm^2 in terms of energy density pulses. In the articles, it was also reported, in particular, that the laser damage threshold of the ZGP surface is limited by the level of the power density of the pump radiation, but not by the radiation intensity [16]. The direct dependence of LIDT on the growth technology and optical quality of crystals was demonstrated in [14].

In [17,18], it was shown that improving the polishing of the ZGP working surfaces and a decrease or complete removal of the near-surface fractured layer leads to an increase in the breakdown threshold. In [17], it was shown that a decrease in the near-surface fractured layer led to a decrease in R_q by two times, a change in the PV parameter by more than five times, and the LIDT with regard to the energy density increased by two times. The LIDT at a wavelength of 2.05 µm and the pulse frequency of 10 kHz for ZGP samples with a sputtered antireflection coating was improved from 1 J/cm^2 to 2 J/cm^2. An increase in LIDT was achieved by improving the polishing of the surface of the ZGP samples. At the same time, the results of the studies presented in [14] show that an almost unchanged polishing parameter R_Z and the variation of the R_q parameter by more than four times, as well as the variation of the R_a parameter by more than five times did not lead to any changes in LIDT. This was assumed to be because the irregularities of the polished surface (peaks and depressions) described by the R_z parameter contribute to the optical breakdown mechanism, and can "seed" inhomogeneities to initialize the optical breakdown due to field effects at a wavelength of 2.091 µm.

Thus, previous studies indicate that the quality of the polishing of the ZGP surface significantly affects the LIDT value, which in turn limits the reliability of the coherent mid-IR radiation sources produced based on a nonlinear ZGP crystal. These circumstances stimulate the development of new polishing methods that provide better surface roughness and a higher LIDT value.

One of the promising methods for the improvement of the surface quality is magnetorheological polishing) [19], which, among other things, is increasingly used in the processing of laser crystals to increase the radiation resistance threshold and reduce the roughness level. The research presented in this article is devoted to checking the possibility of using magnetorheological processing as a method for the removal of a defect layer after fine polishing and the reduction of the level of surface roughness of a single-crystal ZGP. The definition of a defective layer after fine polishing was understood as a fractured layer formed during the mechanical polishing of optical materials. As a rule, the depth of this layer is approximately twice the size of the abrasive used [20]. The influence of magnetorheological processing on the LIDT of ZGP was estimated.

2. Samples under Study and their Parameters

Two samples of a single-crystal ZGP were used for the research—sample No. 1 and No. 2—with dimensions of 6 × 6 × 20 mm^3. The samples under study were cut from a single-crystal ZGP boule (manufactured by LLC "LOK", Tomsk, Russia) at angles θ = 54.5° and φ = 0° relative to the optical axis. A single-crystal ZnGeP$_2$ boule was grown using the Bridgman method in the vertical direction on an oriented seed; the growth was carried out from a molten polycrystalline compound previously synthesized using the two-temperature method [21]. The radiation absorption, taking into account multiple reflections from the crystal faces, at a wavelength of 2.097 µm for both samples, was 0.03 cm^{-1}.

The phase composition of the samples under study was determined prior to the study using X-ray diffraction analysis. According to the result of the X-ray structural analysis, no foreign phases were detected in any the samples under study (Table 1).

Table 1. Results of the X-ray diffraction analysis of the samples under study.

Sample	Detected Phases	Phase Content, Mass %	Lattice Parameters, Å
Sample No. 1	ZnGeP$_2$	100	a = 5.4706 c = 10.7054
Sample No. 2	ZnGeP$_2$	100	a = 5.4707 c = 10.7056

Holograms of the internal volume of the samples under study were obtained using a digital holographic camera DHC-1.064, manufactured by LLC "LOK". The reconstruction of the produced digital holograms was carried out to characterize the volumetric defects. The limiting resolution of the method was 3 µm (a detailed description of the digital holography technique, including those applied to the visualization of defects in ZGP and a description of the holographic camera used, is given in [22]). No volume defects with linear dimensions ≥ the limiting resolution of the applied holographic method were detected in any of the three samples used in this work.

The initial polishing of the working surfaces of both test samples was carried out on a 4-PD-200 polishing and finishing machine (SZOS, Minsk, Belarus). The initial processing of the working surfaces of all of the samples consisted of polishing on a cambric polishing pad using ACM 0.5/0 synthetic diamond powder (with an average grain size of 270 nm). The removal of the material was ~50 µm, which allowed the removal of the fractured layer formed in the process of cutting the crystal into oriented plates, and their preliminary grinding. Then, the samples were additionally polished on a cambric polishing pad using ACM 0.25/0 synthetic diamond powder. After that, the samples were polished on a resin polishing pad made of polishing resin using ACM 0.25/0 synthetic diamond powder.

The working surfaces of sample No. 2 were additionally subjected to magnetorheological processing (MRP) from two ends. The MRP was carried out on a five-axis CNC machine, UMO-00.00.000 (UMO-00.00.000, ITMO, Minsk, Belarus). For the MPR a non-aqueous liquid with magnetic particles of carbonyl iron and nanodiamonds was used. A two-stage MRP was used to increase the productivity of the material removal from the surface, which included hard and soft modes, differing in the size of the gap between the impeller and the workpiece. A ZGP crystal sample was fixed on the installation using a holder made of fluoroplastic.

After the polishing process, the samples were washed in accordance with the conventional method and the MRP technology. Washing was carried out both on the working edges and on the lateral edges, on which glue was applied to fasten the individual elements into blocks. The main contaminants of the polished surface of ZGP were chemical impurities from the water used in the polishing process, residues of polishing resin and polishing powder, remnants of instant glue and picein, as well as fine dust particles. In the first stage, the glued block was placed in high-purity acetone (C_3H_6O) heated to a boiling point of ~50 °C for the gluing. After the gluing, the elements were washed in bidistilled water and placed on a glass support, then lowered into a container with acetone (these operations were performed with the constant irrigation of the crystal surface with bidistilled water to prevent the surface from drying out). Then, the container with acetone was placed in an ultrasonic bath (USB) (JP-030S, Skymen, Shenzhen, China), the frequency of the ultrasonic vibrations was 20 kHz, the power of the piezoelectric element was 50 W. The acetone in the ultrasonic system was heated to the boiling point, and the ultrasound was turned on. The elements in the USB were washed in boiling acetone for 40–50 min. The procedure was repeated 5 times. With each subsequent iteration, the acetone remaining after the process was poured out and pure acetone was poured into the bath. This was performed in order to prevent the contaminating components dissolved in the acetone from re-settling on the crystal surface. When removing the elements from the bath and replacing the acetone, the elements were continuously irrigated with acetone to prevent the crystal surfaces from drying out. Then, the glass holder with the elements was placed in a container with bidistilled water and transferred to an ultrasonic system filled with phosphoric acid (H_3PO_4) diluted with bidistilled water in a ratio of 1:3. Furthermore, the washing was carried out in the USB with heating up to 70 °C, and with a turned-on ultrasound for 10 min. Then, the glass holder with the elements was placed in a container with bidistilled water and transferred to the USB filled with bidistilled water. Washing was carried out in the USB at room temperature, and with a turned-on ultrasound for 10 min. Then, the glass holder with the elements was placed in a container with bidistilled water and transferred to the USB filled with boiling acetone. Washing was carried out with a turned-on ultrasound for 30 min. Then, the glass holder with the elements was placed in a container with bidistilled water and transferred to the USB filled with boiling isopropyl ethanol ($CH_3CH(OH)CH_3$) (70 °C). Then, the glass holder with the elements was placed in a container with bidistilled water, after which the elements were dried. At the final stage of washing, the optical control of the crystal surface was carried out using an optical microscope with 30× magnification. If necessary, fine contamination was removed using cotton swabs soaked in acetone. The washed elements were transported in special membrane containers (so that the working surfaces of the crystal did not touch the walls of the containers). The packing into the shipping containers was carried out in an atmosphere of inert argon gas. The surface roughness of samples 1 and 2 was measured on a 3D optical profilometer, MicroXAM-800 (KLA-Tencor, Milpitas, CA, USA). Sample 2 was measured twice before and after MRP. A PSI phase mode and a Nikon X50 lens (Tokyo, Japan) were used for all of the samples. The field of view was 116 μm × 152 μm. The following parameters were assessed in accordance with ISO 4287-2014: the root-mean-square roughness depth (R_q), the arithmetic mean deviation of the roughness profile from the midline (R_a), and the sum of the average absolute values of the heights of the five largest profile protrusions and the depths of the five largest profile valleys (R_z). The results of measurements of the surface roughness of samples No. 1 and No. 2 are shown in Figures 1–3.

The material loss from the surface after MRP was estimated using the gravimetric method, for which a Pioneer PA214C analytical balance (Ohaus, Parsippany, NJ, USA) with a measurement resolution of 0.0001 g was used. The density taken in the calculations of the material loss from the surface was 4.16 g/cm^3.

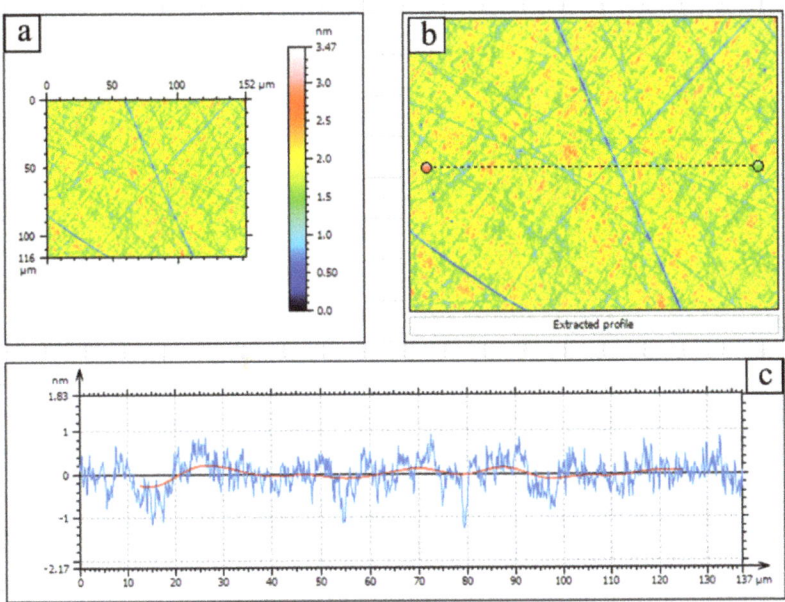

Figure 1. Surface topography and roughness profile of sample No. 1. (**a**) Surface topography of sample, (**b**) Surface topography of sample with a dashed line, (**c**) roughness profile of the sample along the dashed line.

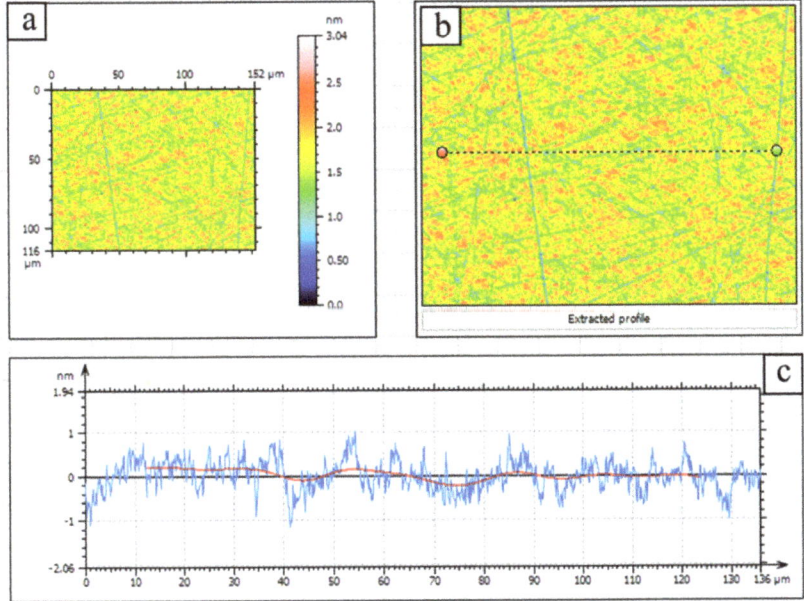

Figure 2. Surface topography and roughness profile of sample No. 2, before MRP. (**a**) Surface topography of sample, (**b**) Surface topography of sample with a dashed line, (**c**) roughness profile of the sample along the dashed line.

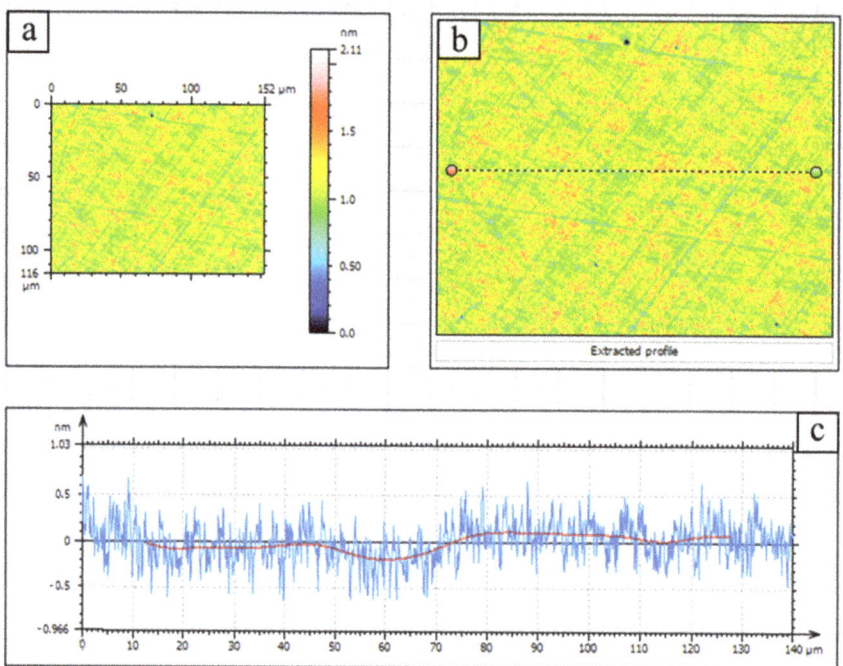

Figure 3. Surface topography and roughness profile of sample No. 2 after MRP. (**a**) Surface topography of sample, (**b**) Surface topography of sample with a dashed line, (**c**) roughness profile of the sample along the dashed line.

3. Setup Parameters and the Technique for the Determination of the LIDT Threshold of the Samples under Study

A Ho:YAG laser generating radiation at a wavelength of 2097 μm, pumped by a continuous thulium fiber laser, was the source of radiation. The Ho:YAG laser operated in the active Q-switched mode with a pulse duration of τ = 35 ns and a pulse repetition rate of 10 kHz. The measured diameter in all of the experiments was d = 350 $\pm$ 10 μm at the e^{-2} level of the maximum intensity. The maximum average radiation power generated by the Ho:YAG laser was 20 W in a linearly polarized Gaussian beam (parameter M2 $\leq$ 1.2).

The schematic layout of the experimental stand is shown in Figure 4. The power of the incident laser radiation was changed using an attenuator consisting of a half-wave plate ($\lambda/2$) and a polarizing mirror (M1). A Faraday isolator (F.I.) was used to prevent the reflected radiation from entering the laser, which prevented an uncontrolled change in the parameters of the incident radiation. The average laser power (Pav) was measured before each experiment with an Ophir power meter (P.M.), 30(150)A-BB-18, Jerusalem, Israel.

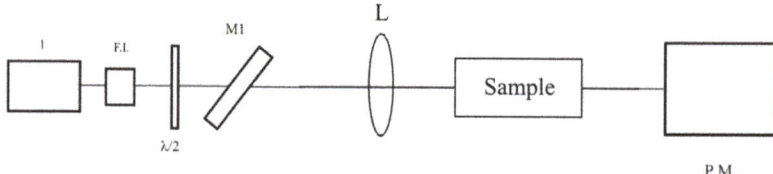

Figure 4. Optical schematic layout of the experimental setup: 1 is the Ho:YAG laser, F.I. is the Faraday isolator, $\lambda/2$ is the half-wave plate, M1 is the polarizing mirror, L is the lens, and P.M. is the Ophir power meter.

According to the international standard ISO11146 [23], the effective area of a Gaussian beam is determined as $S = \pi d^2/4$ [23]. The energy density of the laser radiation was determined by the following equation:

$$W = 8\, P_{av}/(f\pi d^2), \qquad (1)$$

The energy density of the laser irradiation was determined by Equation (2):

$$P = 8\, P_{av}/(f\,\tau\pi d^2) \qquad (2)$$

where d is the diameter of the laser beam, f is the pulse repetition rate, and τ is the duration of the laser pulses.

The "R-on-1" technique was used to determine the LIDT of the samples, which requires less space on the sample surface compared to the "S-on-1" technique, and therefore can be used for samples with a limited aperture; however, it is considered coarser [24]. The essence of this technique is that each individual region of the crystal is irradiated with laser radiation with a sequential increase in the intensity of the laser radiation until an optical breakdown occurs or a predetermined value of the energy density is reached. In our work, the study was carried out with an exposure duration of τ_{ex} = 5 s. The sample under study was exposed to packets of laser pulses with a fixed energy density level, which did not cause damage to the crystal surface. Then, the energy density level was increased with a step of ~0.1 J/cm². The experiment was terminated when visible damage appeared on one of the surfaces of the nonlinear element. Then, the sample was moved 0.5 mm in height or width using a two-dimensional movement; the experiment was repeated five times. The optical breakdown probability was obtained by plotting the cumulative probability versus the optical breakdown energy density. The value of the LIDT (W_{0d}) was taken to be the energy density corresponding to the approximation of the optical breakdown probability to zero. Figure 5 shows the results of the measurement of the LIDT using the R-on-1 technique. In the presented plots, the ordinate represents the probability of optical breakdown in relative units, normalized to unity, and the abscissa represents the energy density of the testing laser radiation.

The average value of the threshold energy density W_{av} and the mean square error of its determination $\langle \Delta W_{av}^2 \rangle$ were calculated for each series of measurements after which an optical breakdown was observed, using the following equations:

$$W_{av} = \frac{\sum W_i n_i}{N} \qquad (3)$$

$$\langle \Delta W_{av}^2 \rangle = \frac{\sum (\langle W_{av}\rangle - W_i)^2 n_i}{N(N-1)} \qquad (4)$$

where N is the total number of damaged areas, W_i is the threshold energy density in one of the irradiated regions, and n_i is the number of regions with a breakdown threshold W_i.

To find the confidence interval of the LIDT value (W_D)

$$W_D = W_{av} \pm k\, \langle \Delta W_{av}^2 \rangle^{1/2}, \qquad (5)$$

where k is Student's coefficient; Student's t-distribution was used for the confidence probability [25,26].

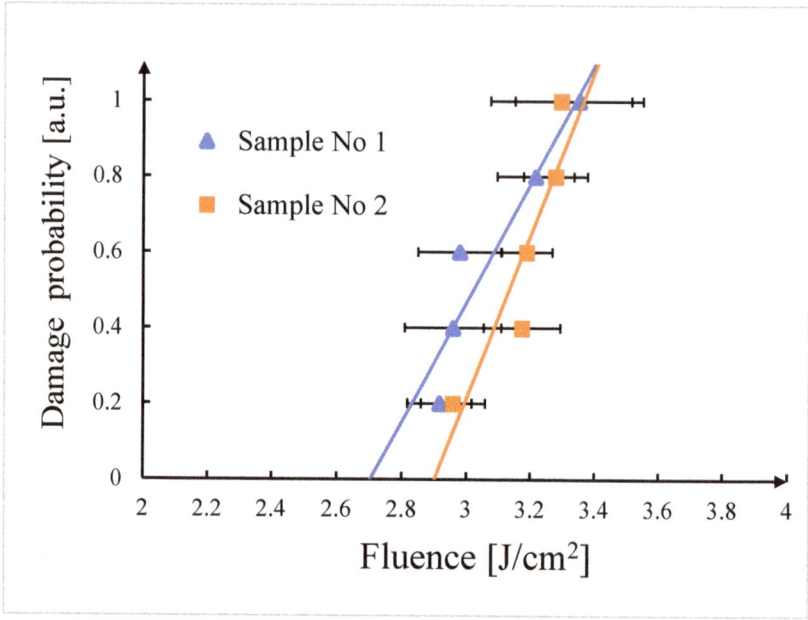

Figure 5. Dependence of the optical breakdown probability of samples No. 1 (Δ) and No. 2 (□) on the energy density of the incident laser radiation.

$$F(k, N) = \frac{\Gamma\left(\frac{N}{2}\right)}{\sqrt{\pi(N-1)}\Gamma\left[\frac{(N-1)}{2}\right]} \int_{-k}^{k} \left(1 + \frac{z^2}{N-1}\right)^{-N/2} dz \qquad (6)$$

where Γ is the gamma function.

After the absorption of the samples was determined, the values of the LIDT were obtained in terms of the energy density, W_{od}^E, and the power density, W_{od}^P, of the testing laser radiation at the probability $P_D = 0$ for each sample, according to the method described above. The average value of the energy density, W_{av}^E, and the power density, W_{av}^P, of the testing radiation, at which the optical breakdown of the sample occurred, was calculated using Equations (1)–(6), and the confidence interval of the LIDT of the values in terms of energy density, W_D^E, and power density, W_D^P, at a confidence level of 0.98 was determined. The experimental results are presented in Table 2.

Table 2. The results of the determination of the LIDT of the studied ZGP samples. The values of the energy density W_{od}^E and the power density W_{od}^P with a probability of the optical breakdown of 0; the average value of the energy density W_D^E and the power density W_D^P, considering the measurement error; Student's coefficient k with a confidence probability of 0.98; and the number of measurements, N.

Sample	N	f, kHz	τ, ns	k	W_D^E, J/cm² (λ-2097 мкм)	W_{od}^E, J/cm² (λ-2097 мкм)	W_D^P, MW/cm² (λ-2097 мкм)	W_{od}^P, MW/cm² (λ-2097 мкм)
No. 1	5	50	35	3.7	(3.1 ± 0.3)	2.7	(88 ± 9)	77
No. 2	5	50	35	3.7	(3.2 ± 0.2)	2.9	(91 ± 6)	83

4. Experimental Results and their Discussion

The results of the measurement of the roughness parameters of the samples are presented in Table 3.

Table 3. Surface roughness parameters of the samples.

Roughness Parameters	Sample 1	Sample 2	
		Initial	after MRP
R_z, nm	1.56	1.46	1.06
R_a, nm	0.227	0.218	0.154
R_q, nm	0.289	0.274	0.193

The analysis of the surface topography of sample No. 1 and sample No. 2, polished using conventional technology, showed that the surface relief was formed under the influence of the multidirectional movement of the working tool; there are single extended scratches up to 1.3 nm deep. The surface topography of sample 2 after MRP did not contain the indicated scratches, and is represented by a less-textured profile formed under the influence of a magnetorheological fluid. A significant improvement in the roughness parameters by 1.37–1.42 times was observed near the surface after MRP. In contrast to [17,18], where the authors reached the nanometer and subnanometer level of the surface roughness of the ZGP crystal samples, the samples studied in this article had an Angstrom roughness level of Ra 2.27 Å for sample No. 1 and Ra 1.54 Å for sample No. 2 after MRP.

All of the surfaces have single point depressions, most likely due to the imperfection of the internal structure of the crystal. These defects are hardly noticeable after traditional polishing because they are partially or completely erased. After MRP, the material is removed from the surface practically without any damage, which more clearly visualizes the areola of the defect and allows the establishment of the true size of the point structural imperfection, i.e., 0.5–1.5 μm.

The removal of the material after MRP from the surfaces of sample 2 was as follows: side A was 6.95 μm and side B was 9.50 μm. In fact, an area of 22 mm × 20 mm instead of 6 mm × 6 mm was treated after MRP. Therefore, the total processing time for side A was 435 min, and for side B it was 345 min. If we subtract the time associated with the acceleration and reversal of the working tool from the total MPR time, then the effective time spent on the working tool on the crystal surface was 8.2% of the total time: for side A it was 28 min, and for side B it was 36 min. In this regard, it is advisable in the future to provide a group type of crystal processing during MRP to increase the efficiency of the use of the equipment.

Figures 5 and 6, and Table 2 show the results of the LIDT study of sample No. 1 (polished using conventional technology) and sample No. 2 immediately after MRP polishing.

As can be seen from the results of the determination of the LIDT (Figures 5 and 6, and Table 2), the difference in the LIDT in the energy density and the power density for two samples fits into the error of the LIDT determination technique, even though a significant improvement in the roughness parameters by 1.37–1.42 times was observed on the surfaces of sample No. 2 after MRP. At first, the results obtained contradicted the data [17,18], in which an improvement in roughness parameters by 2.1 times led to an increase in LIDT by 1.6 times. However, it should be noted that, in [17], the R_z parameter was reduced from 225 nm to 41 nm, and the R_q parameter was reduced from 1.2 nm to 0.57 nm (the minimum achieved value of the R_z parameter was an order of magnitude larger than the dimension of the parameters of the ZGP crystalline lattice). Based on the results presented in Table 3, both polishing techniques presented in this article allowed us to obtain a surface with a roughness (estimated by the parameters R_a, R_q, and R_z) of the same order of dimension with the parameters of the unit cell of the ZGP crystalline lattice (a = b = 0.547 nm and c = 1.07 nm, Table 1).

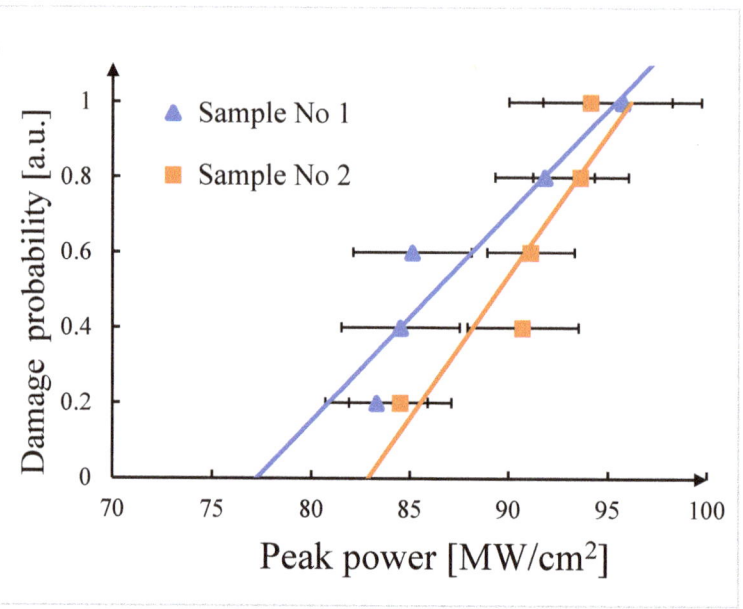

Figure 6. Dependence of the optical breakdown probability of samples No. 1 (Δ) and No. 2 (□) on the power density of the incident laser radiation.

The absence of a distinguishable difference in the breakdown threshold for the two samples is most likely conditioned upon the fact that the LIDT value at the indicated orders of magnitude of the surface roughness parameters is determined not by the quality of polishing, but by the number of point depressions caused by the physical limitations of the structural configuration of the crystal volume. These results are in good agreement with the assumption made in [14] about the significant effect of the concentration of dislocations in a ZGP crystal on LIDT. It was noted in [27] that at a sufficiently low concentration, zero-dimensional defects (dislocations, bulk defects), practically without affecting the conditions of radiation propagation due to a weak shadow effect, can significantly reduce the LIDT at the points of their emergence on the surface, playing the role of "nuclei" (or "weak links"), from which the irreversible process of the matrix crystal destruction begins under the action of optical beams of extreme intensity. It can be assumed that the observed point depressions, which are clearly distinguishable during MRP polishing—having dimensions of 0.5–1.5 µm—are a consequence of the "emergence" of dislocations on the surface during the polishing process. As is known, mechanically stressed regions emerge around the dislocations, the presence of which in the process of polishing can cause the appearance of detected point depressions with the above dimensions. An alternative explanation of the results obtained is the presence of volume defects 0.5–1.5 µm in size in the studied ZGP crystals due to the presence of impurities of the intrinsic components in the initial synthesized material, in the form of binary phosphides of zinc and germanium (Zn_3P_2, ZnP_2, GeP). These defects could not have been detected during the characterization of the samples under study, as the resolution of the digital holographic camera used to detect volumetric defects is 3 µm, which is half the size of the detected irregularities. The detective power of the X-ray structural analysis may also have turned out to be insufficient. From the above, it can be assumed that some of the possible ways to increase the LIDT are to further improve the technology of synthesis and crystal growth, to minimize defects in the crystal structure, and to improve the characteristics of the optical coatings.

5. Conclusions

Samples of a single-crystal ZnGeP$_2$ with an angstrom level of surface roughness were investigated in this article. Magnetorheological processing was applied to polish the working surfaces of the single-crystal ZGP, in which a non-aqueous liquid with magnetic particles of carbonyl iron with the addition of nanodiamonds was used. The material showed good polishability: MRP led to a significant improvement in the surface roughness parameters by 1.37–1.42 times (R_a 1.54 Å), compared to the conventional crystal polishing technique (R_a 2.27 Å) using an aqueous suspension based on diamond powder and a resin pad. The removal of the material from the crystal surface after MRP ranged from 6.95 to 9.5 µm. Moreover, the use of MRP allowed the more accurate characterization of possible structural defects that emerge on the surface of a single crystal and have a size of ~0.5–1.5 µm. The useful processing time of the 6 × 6 sample was 8.2% of the total processing time. It is recommended to use the group type of crystal processing during MRP in order to minimize the time spent on the idling and reverse of the working tool, which will significantly increase the efficiency of the use of the industrial equipment. Thus, both polishing methods allowed us to obtain an Angstrom level of surface roughness comparable in its order of magnitude with the unit cell parameters of the ZGP crystalline lattice, which indicates that the surface quality after MRP is close to the maximum possible.

Despite the fact that the sample subjected to MRP showed a significant improvement in the surface roughness parameters compared to the sample polished using the conventional technology, the LIDT remained practically unchanged. The absence of a difference in the breakdown threshold for the two samples was most likely due to the fact that the LIDT value at Angstrom parameters of surface roughness is determined to a greater extent not by the quality of polishing, but by the physical limitations of the structural configuration of the crystal. It has been suggested that the LIDT is most influenced by dislocations or volume defects "emerging" on the polished surface, rather than by the roughness level. Thus, at the Angstrom level of roughness, the decisive factor affecting the LIDT value is the concentration of bulk defects "emerging" on the crystal surface.

Author Contributions: Conceptualization, N.Y. and A.K.; methodology, M.Z.; software, G.G. and P.K.; validation, N.Y. and E.S.; formal analysis, O.A.; investigation, N.Y., M.Z., S.P., A.K., G.G. and P.K.; resources, N.Y. and A.K.; data curation, E.S. and E.Z.; writing—original draft preparation, E.S.; writing—review and editing, N.Y. and S.P.; visualization, M.K.; supervision, N.Y. and A.K.; project administration, N.Y. and A.K.; funding acquisition, N.Y. All authors have read and agreed to the published version of the manuscript.

Funding: This study was supported by the Tomsk State University Development Programme («Priority-2030»).

Data Availability Statement: not applicable.

Conflicts of Interest: The authors declare no conflict of interest. The funders had no role in the design of the study; in the collection, analyses, or interpretation of data; in the writing of the manuscript, or in the decision to publish the results.

References

1. Yevtushenko, A.; Rozniakowska-Klosinska, M. *Encyclopedia of Thermal Stresses, Laser-Induced Thermal Splitting in Homogeneous Body with Coating*; Springer: Berlin, Germany, 2014.
2. Parfenov, V.A. *Laser Materials Microprocessing*; SPbGETU "LETI": Saint-Petersburg, Russia, 2011; p. 59.
3. Kozub, J.; Ivanov, B.; Jayasinghe, A.; Prasad, R.; Shen, J.; Klosner, M.; Heller, D.; Mendenhall, M.; Piston, D.W.; Joos, K.; et al. Raman-shifted alexandrite laser for soft tissue ablation in the 6- to 7-µm wavelength range. *Biomed. Opt. Express* **2011**, *2*, 1275–1281. [CrossRef]
4. Bobrovnikov, S.M.; Matvienko, G.G.; Romanovsky, O.A.; Serikov, I.B.; Sukhanov, A.Y. *Lidar Spectroscopic Gas Analysis of the Atmosphere*; IOA SB RAS: Tomsk, Russia, 2014; p. 510.
5. Romanovskii, O.A.; Sadovnikov, S.A.; Kharchenko, O.V.; Yakovlev, S.V. Development of Near/Mid IR differential absorption OPO lidar system for sensing of atmospheric gases. *Opt. Laser Technol.* **2019**, *116*, 43–47. [CrossRef]
6. Bochkovskii, D.A.; Vasil'eva, A.V.; Matvienko, G.; Yakovlev, S.V. Application of a strontium vapor laser to laser remote sounding of atmospheric composition. *Atmos.Ocean. Opt.* **2012**, *25*, 166–170. [CrossRef]

7. Schunemann, P.G.; Zawilski, K.T.; Pomeranz, L.A.; Creeden, D.J.; Budni, P.A. Advances in nonlinear optical crystals for mid-infrared coherent sources. *J. Opt. Soc. Am. B* **2016**, *33*, D36–D43. [CrossRef]
8. Hemming, A.; Richards, J.; Davidson, A.A.; Carmody, N.; Bennetts, S.; Simakov, N.; Haub, J. 99 W mid-IR operation of a ZGP OPO at 25% duty cycle. *Opt. Express* **2013**, *21*, 10062–10069. [CrossRef] [PubMed]
9. Haakestad, M.W.; Fonnum, H.; Lippert, E. Mid-infrared source with 0.2 J pulse energy based on nonlinear conversion of Q-switched pulses in $ZnGeP_2$. *Opt. Express* **2014**, *22*, 8556–8564. [CrossRef] [PubMed]
10. Qian, C.; Yao, B.; Zhao, B.; Liu, G.; Duan, X.; Ju, Y.; Wang, Y. High repetition rate 102 W middle infrared $ZnGeP_2$ master oscillator power amplifier system with thermal lens compensation. *Opt. Lett.* **2019**, *44*, 715–718. [CrossRef] [PubMed]
11. Hildenbrand, A.; Kieleck, C.; Tyazhev, A.; Marchev, G.; Stöppler, G.; Eichhorn, M.; Schunemann, P.G.; Panyutin, V.L.; Petrov, V. Laser damage of the nonlinear crystals $CdSiP_2$ and $ZnGeP_2$ studied with nanosecond pulses at 1064 and 2090 nm. *Opt. Eng.* **2014**, *53*, 122511. [CrossRef]
12. Gribenyukov, A.I.; Dyomin, V.V.; Olshukov, A.S.; Podzyvalov, S.N.; Polovcev, I.G.; Yudin, N.N. Investigation of the process of laserinduced damage of $ZnGeP_2$ crystals using digital holography. *Russ. Phys. J.* **2018**, *61*, 2042–2052. [CrossRef]
13. Yudin, N.N.; Zinoviev, M.; Gladkiy, V.; Moskvichev, E.; Kinyaevsky, I.; Podzyvalov, S.; Slyunko, E.; Zhuravleva, E.; Pfaf, A.; Yudin, N.A.; et al. Influence of the Characteristics of Multilayer Interference Antireflection Coatings Based on Nb, Si, and Al Oxides on the Laser-Induced Damage Threshold of $ZnGeP_2$ Single Crystal. *Crystals* **2021**, *11*, 1549. [CrossRef]
14. Yudin, N.N.; Antipov, O.L.; Gribenyukov, A.I.; Eranov, I.D.; Podzyvalov, S.N.; Zinoviev, M.M.; Voronin, L.A.; Zhuravleva, E.V.; Zykova, M.P. Effect of postgrowth processing technology and laser radiation parameters at wavelengths of 2091 and 1064 nm on the laser-induced damage threshold in $ZnGeP_2$ single crystal. *Quantum Electron.* **2021**, *51*, 306–316. [CrossRef]
15. Andreev, Y.M.; Badikov, V.V.; Voevodin, V.G.; Geiko, L.G.; Geiko, P.P.; Ivashchenko, M.V.; Karapuzikov, A.I.; Sherstov, I.V. Radiation resistance of nonlinear crystals at a wavelength of 9.55 μm. *Quantum Electron.* **2001**, *31*, 1075–1078. [CrossRef]
16. Peterson, R.D.; Schepler, K.L.; Brown, J.L.; Schunemann, P.G. Damage properties of $ZnGeP_2$ at 2 μm. *J. Opt. Soc. Am. B* **1995**, *12*, 2142–2146. [CrossRef]
17. Zawilski, K.T.; Setzler, S.D.; Schunemann, P.G.; Pollak, T.M. Increasing the laser-induced damage threshold of single-crystal $ZnGeP_2$. *J. Opt. Soc. Am. B* **2006**, *23*, 2310–2316. [CrossRef]
18. Lei, Z.; Zhu, C.; Xu, C.; Yao, B.; Yang, C. Growth of crack-free $ZnGeP_2$ large single crystals for high-power mid-infrared OPO applications. *J. Cryst. Growth* **2014**, *389*, 23–29. [CrossRef]
19. Sutowska, M.; Sutowski, P. Contemporary applications of magnetoreological fluids for finishing process. *J. Mech. Energy Eng.* **2017**, *1*, 141–152.
20. Gerhard, C.; Stappenbeck, M. Impact of the polishing suspension concentration on laser damage of classically manufactured and plasma post-processed zinc crown glass surfaces. *Appl. Sci.* **2018**, *8*, 1556. [CrossRef]
21. Verozubova, G.A.; Gribenyukov, A.I.; Mironov, Y.P. Two-temperature synthesis of $ZnGeP_2$. *Inorg. Mater.* **2007**, *43*, 1040–1045. [CrossRef]
22. Dyomin, V.; Gribenyukov, A.; Davydova, A.; Zinoviev, M.; Olshukov, A.; Podzyvalov, S.; Polovtsev, I.; Yudin, N. Holography of particles for diagnostics tasks [Invited]. *Appl. Opt.* **2019**, *58*, G300–G310. [CrossRef] [PubMed]
23. *Lasers and Laser-Related Equipment–Test Methods for Laser Beam Widths, Divergence Angles and Beam Propagation Ratios*, 1st ed.; ISO: Geneva, Switzerland, 2005.
24. "The R-on-1 Test," Lidaris LIDT Service. 2019. Available online: http://lidaris.com/laserdamage-testing/ (accessed on 26 November 2021).
25. *Statistical Interpretation of Test Results–Estimation of the Mean–Confidence Interval*, 1st ed.; ISO: Geneva, Switzerland, 1980.
26. Fisher, R.A.; Rothamsted, M.A. statistical methods for research workers. *Metron* **1925**, *5*, 90.
27. Verozubova, G.A.; Filippov, M.M.; Gribenyukov, A.I.; Trofimov, A.Y.; Okunev, A.O.; Stashenko, V.A. Investigation of the evolution of structural defects in $ZnGeP_2$ single crystals, grown by the bridgman method. *Math. Mech. Phys.* **2012**, *321*, 121–128.

Article

Influence of the Characteristics of Multilayer Interference Antireflection Coatings Based on Nb, Si, and Al Oxides on the Laser-Induced Damage Threshold of ZnGeP$_2$ Single Crystal

Nikolai Nikolayevich Yudin [1,2,3,*], Mikhail Zinoviev [1,2,3], Vladislav Gladkiy [4], Evgeny Moskvichev [1], Igor Kinyaevsky [5], Sergey Podzyvalov [1,3], Elena Slyunko [1,3], Elena Zhuravleva [1], Anastasia Pfaf [1], Nikolai Aleksandrovich Yudin [1,2] and Maxim Kulesh [1,*]

[1] Laboratory for Radiophysical and Optical Methods of Environmental Studies, National Research Tomsk State University, 634050 Tomsk, Russia; muxa9229@gmail.com (M.Z.); em_tsu@mail.ru (E.M.); cginen@yandex.ru (S.P.); elenohka266@mail.ru (E.S.); lenazhura@mail.ru (E.Z.); nastya.pfayff17@mail.ru (A.P.); yudin@tic.tsu.ru (N.A.Y.)

[2] Laboratory of Remote Sensing of the Environment, V.E. Zuev Institute of Atmospheric Optics SB RAS, 634055 Tomsk, Russia

[3] Department of Optical Measurements, LOC LLC, 634050 Tomsk, Russia

[4] STC IZOVAC, 220075 Minsk, Belarus; info@izovac.com

[5] Laboratory of Gas Lasers, P.N. Lebedev Institute of Physics RAS, 119333 Moscow, Russia; kinyaevskiyio@lebedev.ru

* Correspondence: rach3@yandex.ru (N.N.Y.); kyleschMM2000@yandex.ru (M.K.); Tel.: +7-996-938-71-32 (N.N.Y.); +7-983-237-90-48 (M.K.)

Abstract: In this work, the effect of the defect structure and the parameters of antireflection interference coatings based on alternating layers of Nb$_2$O$_5$/Al$_2$O$_3$ and Nb$_2$O$_5$/SiO$_2$ layers on the laser-induced damage threshold of ZGP crystals under the action of Ho:YAG laser radiation at a wavelength of 2.097 µm was determined. Coating deposition was carried out using the ion-beam sputtering method. The laser-induced damage threshold of the sample with a coating based on alternating layers Nb$_2$O$_5$ and SiO$_2$ was $W_{0d} = 1.8$ J/cm^2. The laser-induced damage threshold of the coated sample based on alternating layers of Nb$_2$O$_5$ and Al$_2$O$_3$ was $W_{0d} = 2.35$ J/cm^2. It has been found that the presence of silicon conglomerates in an interference antireflection coating leads to a decrease in the laser-induced damage threshold of a nonlinear crystal due to local mechanical stresses and the scattering of incident laser radiation.

Keywords: laser-induced damage threshold; ZnGeP$_2$; interference coating

1. Introduction

Increasing the power and efficiency of mid-IR laser systems remains one of the main problems of modern laser physics and technology, which is conditioned upon the need for such systems when solving many scientific and applied tasks. Such tasks include remote gas analysis and the monitoring of the atmosphere [1], material processing [2], the investigation of new physical effects [3], generation of attosecond X-ray pulses and particle acceleration [4], and many others. One of the main methods of producing high-power laser radiation in the mid-IR range is the parametric frequency conversion of near-IR lasers [5,6] using nonlinear optical crystals. One of the most effective crystals for this task in the wavelength range of 3–5 µm is a single crystal ZnGeP$_2$ (ZGP) [6,7], which is sometimes called the "standard" of nonlinear crystals in the mid-IR range [6]. ZGP is a nonlinear positive (ne > no) uniaxial crystal with a crystalline lattice of the Chalcopyrite type [8,9]. ZGP has a high value of nonlinear susceptibility of 75 × 10^{-12} m/V and a thermal conductivity of 36 W/(m × K), as compared to other nonlinear crystals [8,9]. However, a high refractive index of ~3 and, consequently, a high degree of pump radiation reflection at the crystal–air interface significantly reduce the efficiency of nonlinear conversion in the crystal. To increase this significantly, it is necessary to apply antireflection coatings on the working surfaces of the crystal in the required conversion range. Strict requirements on the value of optical strength are imposed for such coatings, which, at least, should not be lower than the optical strength of the crystal itself. It should be noted that a reliable long-term operation of powerful nonlinear ZGP-based converters is limited by the optical breakdown effect [3–6]. In this regard, the potential for the practical use of high-power parametric oscillators in the mid-IR range while pumping by radiation in the wavelength range of ~2.1 μm is associated with the need to determine technological factors affecting the laser-induced damage threshold (LIDT). The problem of ZGP optical breakdown by laser radiation at wavelengths from 1.064 μm to 10 μm has been discussed in several previously published works [10–16]. These studies revealed a significant difference in the value of the LIDT of the ZGP crystal at wavelengths of 1064 nm and 2100 nm [10]. A dynamic visualization of the breakdown process by laser radiation at a wavelength of 2100 nm in the ZGP volume showed that an avalanche-like increase in temperature occurs in the forming track inside the nonlinear optical element [11]. An increase in the ZGP breakdown threshold reported in [12] with a decrease in the pump pulse duration "testifies in favor of the thermal nature of breakdown for nanosecond pulses due to anomalous infrared absorption". It was shown in [15] that when the crystal is cooled to a temperature of −60 °C, the LIDT increases by three times up to 9 J/cm^2 at a wavelength of the acting laser radiation of 2.091 μm and a pulse repetition rate of 10 kHz.

In [13], it was reported that the LIDT of ZGP elements at a wavelength of 9.55 μm was determined from the intensity of the incident beam of 142 MW/cm^2 with a pulse duration of 85 ns and a pulse repetition rate of 1 Hz, which is ~9.5 J/cm^2 by the pulse energy density. It was also reported in these works that the threshold for the laser destruction of the ZGP surface is associated with the level of the energy density of the pump radiation rather than the radiation intensity [14]. The direct dependence of the LIDT on the growth technology and the optical quality of crystals was demonstrated in [15]. In [16], it was shown that improving the polishing of working surfaces and reducing or completely removing the near-surface fractured layer leads to an increase in the breakdown threshold. It was shown in [17] that the threshold values of laser damage measured for samples with an antireflection coating are significantly lower than for samples without a coating. At the same time, it was shown in [16] that the deposition of antireflection interference coatings leads to a twofold increase in the LIDT. In [17], it is directly concluded that further research should be focused on improving the quality of the antireflection coating using a high-quality ZGP crystal in order to increase the output power and the efficiency of parametric conversion in a ZGP crystal in the spectral range of 3–5 μm. Thus, the large scatter in the values of the LIDT of the ZGP crystal and the difference in interpretations of the results of studies given in previous works show that physical mechanisms of this negative effect and its dependence—in particular, the technology of deposition of interference coatings—remain not fully understood.

Currently, there are many film-forming materials that can cover a wide range from UV to mid-IR. Several materials, such as Ta_2O_5, Nb_2O_5, SiO_2, HfO_2, YbF_3, ZnS, Al_2O_3, etc., have high transparency and low absorption coefficients in the mid-IR region of the spectrum from 2 to 13 μm [18–22]. These materials are being actively studied and are the most promising as film-forming materials for the development of antireflection coatings and dielectric mirrors with high radiation resistance. Thus, in [18], a dielectric mirror was produced on a fused silica substrate with a radiation strength of about 42 J/cm^2, which is practically commensurate with the strength of the substrate itself. In [19], the effect of the substrate temperature during the deposition of a film-forming material was investigated, and the dependence of the LIDT of the coating depending on the deposition conditions was shown. The authors of [20,21] also show differences in the strength of the produced dielectric coatings depending on the conditions of material deposition. However, most of these studies were carried out using quartz, sapphire substrates, and similar optical windows that are transparent in the mid-IR range. There is practically no information related to the study of the mentioned materials deposited using the method of ion-beam sputtering on crystalline materials of the ZGP type. Thus, it is necessary to study the influence of the deposition parameters of the material and the choice of the material on the optical strength of these coatings.

The contradictory information on the influence of interference coatings on the LIDT of ZGP presented in [16,17] indicates that the different structures of coatings and methods of their deposition on a nonlinear crystal, as well as the choice of the film-forming materials, significantly affects the efficiency of its application in optical parametric oscillators.

The aim of the studies presented in this work is to determine the effect of the defect structure and parameters of antireflection interference coatings on the LIDT of ZGP crystals under the action of Ho:YAG laser radiation at a wavelength of 2.097 μm (the most common pump source for parametric oscillators of mid-IR light based on crystals ZGP). Two types of coatings were investigated: based on alternating layers of Nb_2O_5 and Al_2O_3 and based on alternating layers of Nb_2O_5 SiO_2 layers. Coating deposition was carried out using the ion-beam sputtering method.

2. ZGP Substrates (Manufactured by LLC "LOC", Tomsk, Russia) and Their Parameters

Three samples of a ZGP single crystal were used for studies of the optical breakdown: samples with dimensions of 6 × 6 × 20 mm^3 were cut from a single crystal ZGP boule (manufactured by LLC "LOC", Tomsk, Russia) at the angles of θ = 54.5° and φ = 0° relative to the optical axis. A single crystal $ZnGeP_2$ boule was grown using the Bridgman method in the vertical direction on an oriented seed; the growth was carried out from a molten polycrystalline compound previously synthesized using the two-temperature method [15].

One sample was tested without applying an antireflection coating (for convenience, hereinafter referred to as ZGP_clean), the second sample was coated with an antireflection coating based on alternating Nb_2O_5/Al_2O_3 layers (for convenience, hereinafter referred to as ZGP_Al_2O_3), the third sample was coated with an antireflection coating based on six alternating layers of Nb_2O_5/SiO_2 (for convenience, referred to asZGP_1_SiO_2 hereafter), and the fourth sample was coated with an antireflection coating based on four layers of Nb_2O_5/SiO_2 (for convenience, referred to ZGP_2_SiO_2 hereafter).

The absorption of radiation at a wavelength of 2.097 μm for sample ZGP_clean was 0.029 cm^{-1}, for sample ZGP_Al_2O_3 was 0.03 cm^{-1}, for sample ZGP_1_SiO_2 was 0.028 cm^{-1}, and for sample ZGP_2_SiO_2 was 0.029 cm^{-1}.

The phase composition of the samples under study was determined prior to the study using X-ray diffraction analysis. According to the result of X-ray structural analysis, no foreign phases were detected in all the samples under study (Table 1).

Table 1. Results of X-ray diffraction analysis of the samples under study.

Sample	Detected Phases	Phase Content, Mass %	Lattice Parameters, Å
ZGP_clean	$ZnGeP_2$	100	a = 5.4706 c = 10.7054
ZGP_Al_2O_3	$ZnGeP_2$	100	a = 5.4707 c = 10.7056
ZGP_1_SiO_2	$ZnGeP_2$	100	a = 5.4707 c = 10.7053
ZGP_2_SiO_2	$ZnGeP_2$	100	a = 5.4706 c = 10.7055

Holograms of the internal volume of the samples under study were obtained using a digital holographic camera DHC-1.064, manufactured by LLC "LOC". The reconstruction of the produced digital holograms was carried out in order to characterize volumetric defects. The limiting resolution of the method was 3 μm (a detailed description of the digital holography technique, including those applied to visualization of defects in ZGP and a description of the holographic camera used, is given in [22]). No volume defects were detected in all three samples used in this work.

The working surfaces of the test samples were polished on a 4-PD-200 polishing-and-lapping machine. The initial treatment of the working surfaces of all samples consisted of polishing on a cambric polishing pad using ACM 0.5/0 synthetic diamond powder (average grain size of 270 nm). The amount of material removed was ~50 μm, which allowed the fractured layer formed in the process of cutting the crystal into oriented plates and their preliminary grinding to be removed. Then, the samples were additionally polished on a cambric polishing pad using ACM 0.25/0 synthetic diamond powder. After that, the samples were polished on a resin polishing pad made of polishing resin using ACM 0.25/0 synthetic diamond powder. The control of the profile of the working surfaces of the samples under study before the deposition of interference coatings was carried out using a ZYGO NewView 7300 profilometer, which operates on the basis of white light interferometry (Figure 1). Surface areas with a size of 110 μm × 110 μm were investigated for each experimental sample, and the following parameters were assessed: the maximum difference in the height and the depth of irregularities on the surface (PV) and root mean square depth of roughness (RMS). These were as follows: for sample ZGP_clean, PV = 31.372 nm, RMS = 0.442 nm; for sample ZGP_Al$_2$O$_3$, PV = 34.67 nm, RMS = 0.48 nm; for sample ZGP_1_SiO$_2$ PV = 39.467 nm, RMS = 0.44 nm; and for sample ZGP_2_SiO$_2$, PV = 33.871 nm, RMS = 0.46 nm. Figure 1 shows an image of the surface of sample ZGP_clean as an example, produced using a ZYGO NewView 7300 profilometer. As can be seen from Figure 1, the polished surfaces of the samples have a low roughness, but they contain "cavities" whose depth reaches several tens of nm. We assume that these surface defects are not caused by the polishing process but are caused by the emergence of bulk defects of the ZGP crystal itself on the surface; for example, these defects can be caused by dislocations.

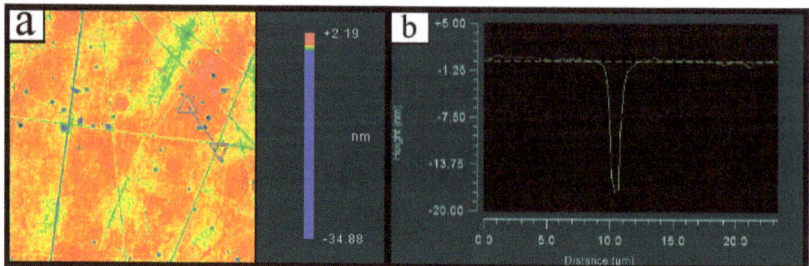

Figure 1. Surface profile of the studied ZGP samples, (**a**) color diagram of surface roughness, (**b**) deviations from the average roughness profile along the line (marked with triangles in (**a**)).

3. Parameters of Interference Coatings Deposition on ZGP Substrates and Their Characteristics

In our work, we used the method of ion-beam sputtering (IBS) of a pure material target with the supply of a reactive gas (oxygen) through a sputtering source. The deposition of dielectric layers was carried out on an Aspira-200 vacuum deposition machine (manufactured by Izovak, Belarus). The sputtered targets were disks of pure material, Si with a purity of 99.999 (5N), Nb with a purity of 99.95 (3N), and Al with a purity of 99.999 (5N). The diameter of each sputtered target was 101.6 mm, and the thickness of the target disk was 6 mm. The targets were fixed on a water-cooled rotary base. The positive charge on the target generated by the ion beam was compensated by the thermal emission of electrons with a hot tungsten cathode. The working gases in the system were gaseous argon (Ar) with a high purity of 99.995% and technical gaseous oxygen (O_2) with a purity of 99.7%.

Before the deposition process, the substrates were preliminarily cleaned from all kinds of dirt and dust on the surface. Cleaning was carried out using phosphoric acid followed by rinsing with high purity acetone. Immediately before coating in a vacuum chamber, the substrates were additionally cleaned with an auxiliary ion source at a source power of about 100 W and an ion energy of about 600 eV for 3 min.

A preliminary evacuation of the vacuum chamber before the beginning of the spraying process was carried out to values of $5 \cdot 10^{-4}$ Pa using a turbomolecular pump. The working residual pressure in the chamber during layer deposition was as follows: for the Nb_2O_5 layer, it was $5 \cdot 10^{-2}$ Pa; for the SiO_2 layer, it was $3.9 \cdot 10^{-2}$ Pa; and for the Al_2O_3 layer, it was $4.8 \cdot 10^{-2}$ Pa. The average deposition rate of the layers was 0.75 A/s for Nb_2O_5, 2 A/s for the SiO_2 layer, and 0.85 A/s for the Al_2O_3 layer.

The primary step for the development and deposition of an interference antireflection coating is to obtain information on the dispersion of the refractive index and absorption in the monolayers from which this coating will be constructed. Thus, we have carried out studies to obtain the optical characteristics of the films of these materials. The thickness of the monolayers was about 1 μm for their correct description in the IR and visible spectral regions. The deposition thickness control was single-wavelength and optical, with a selected wavelength of 550 nm. The monolayers were deposited on pure silicon substrates (refractive index ~3.4) and Asahi optical glass.

After deposition, the monolayers were measured for transmission and reflection on a Shimadzu UV-3600 Plus spectrophotometer (operating wavelength range 180–3300 nm) and a Simex FTIR spectrometer. A typical transmission spectrum of a monolayer is shown in Figure 2 (measurements carried out for Al_2O_3).

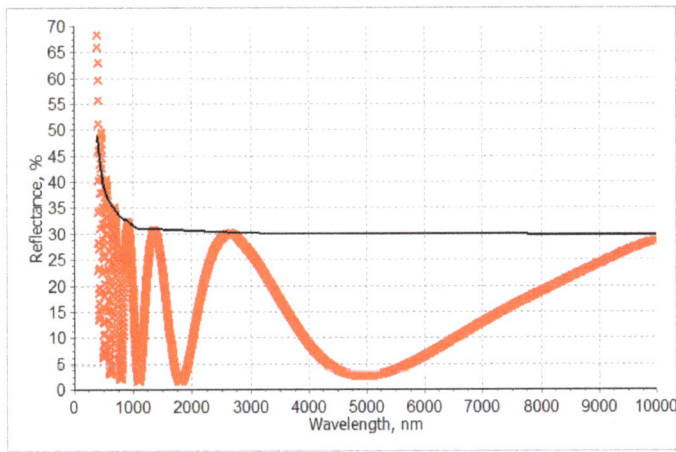

Figure 2. Transmission spectrum of a Al_2O_3 monolayer with a thickness of 1 μm deposited onto a substrate of Asahi optical glass.

The calculation of the dependence of the refractive index and the absorption coefficient on the wavelength of the monolayer was carried out using the Optilayer software and the built-in Optichar module. The obtained dispersions of the refractive index of the monolayers, as well as the dimensionless absorption coefficient, are shown in Figure 3.

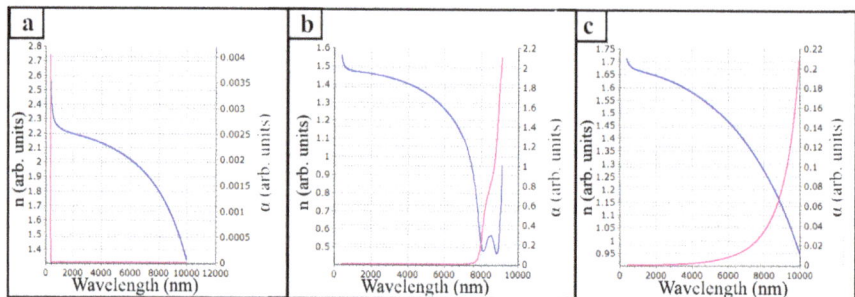

Figure 3. Calculated Optilayer dependences of the refractive index and the absorption coefficient on the wavelength: (**a**) Nb_2O_5 monolayer, (**b**) SiO_2 monolayer, (**c**) Al_2O_3 monolayer.

The obtained dependences of the change in the refractive index and the absorption coefficient on the wavelength showed that all the monolayers under study are transparent in the required range, and the absorption in these layers is insignificant (in the case of Al_2O_3) or is completely absent (as in the case of the Nb_2O_5 and SiO_2 layers). As can be seen from the obtained dispersions, the Al_2O_3 and SiO_2 layers have resonant absorption lines with a sharp jump-like change in the refractive index. For the SiO_2 layer, it begins at a wavelength of ~8 μm and has a steeper rise in the absorption curve, while the Al_2O_3 layer demonstrates a gentler increase in the absorption coefficient with a maximum in the region of 10–12 μm. However, both films have insignificant absorption in the required wavelength range (25 μm) and are suitable as materials for the development of AR coatings. The refractive indices of the produced films are close to the values obtained by the authors of other works. Thus, the refractive index at a wavelength of 550 nm of the Nb_2O_5 film was 2.34, for SiO_2 was 1.47, and for Al_2O_3 was 1.68. According to the literature data, such indicators are typical for stoichiometric close-packed films without significant defects and a porous structure.

Thus, the calculated dispersions of the refractive index and the absorption coefficient of the monolayers under study were subsequently used to design two different AR coatings based on pairs of layers N_2O_5/SiO_2 and Nb_2O_5/Al_2O_3. The development of the coatings was carried out using the Optichar software. The sprayed coating was inspected using an Asahi optical glass witness at a wavelength of 550 nm.

An antireflection coating was not applied to sample ZGP_clean. Sample ZGP_Al$_2$O$_3$ was coated with a four-layer antireflection coating based on compounds Nb_2O_5 (high refractive layer) and Al_2O_3 (low refractive layer) with a thickness of 2133 nm. Sample ZGP_1_SiO$_2$ was coated with a six-layer antireflection coating based on alternating layers of Nb_2O_5 and SiO_2 compounds with a total thickness of ~2900 nm. Sample ZGP_2_SiO$_2$ was coated with a four-layer antireflection coating based on alternating layers of Nb_2O_5 and SiO_2 compounds with a total thickness of ~700 nm. Figure 4 shows the reflection spectra of samples ZGP_Al$_2$O$_3$, ZGP_1_SiO$_2$ and ZGP_2_SiO$_2$ with applied interference coatings, obtained using a Simex Fourier spectrometer. The morphology of interference coatings and the composition of the samples under study were analyzed using a scanning electron microscope with a Tescan MIRA 3 LMU Schottky cathode (TESCAN ORSAY HOLDING, Brno, Czech Republic) equipped with an Oxford Instruments Ultim Max 40 energy dispersive X-ray spectrometer (Oxford Instruments, High Wycombe, UK). Scanning was performed at an accelerating voltage (HV) of 20 kV. The samples were coated with a carbon-conductive coating in a Quorum Technologies EMITECH K450X setup (Quorum Technologies, Laughton, UK). As studies have shown, silicon conglomerates of submicron size 3 were found in sample ZGP_1_SiO$_2$ in the layers of the interference coating (Figure 5b). No defects in the interference antireflection coating were found in sample ZGP_Al$_2$O$_3$.

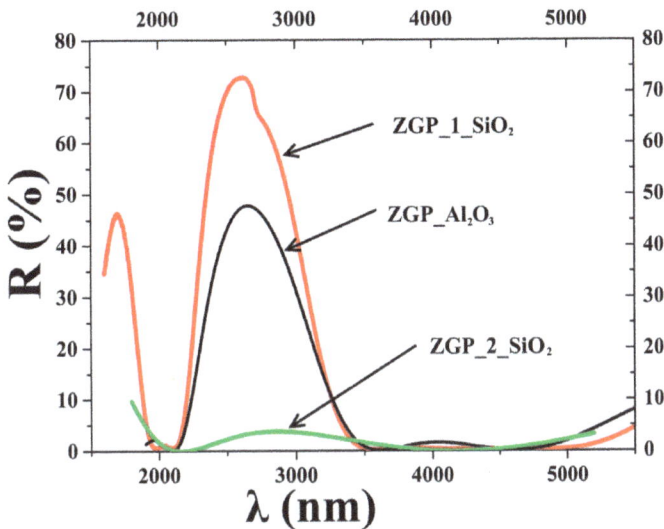

Figure 4. Reflection spectra of working faces of samples ZGP_Al$_2$O$_3$, ZGP_1_SiO$_2$ and ZGP_2_SiO$_2$ with applied interference coatings.

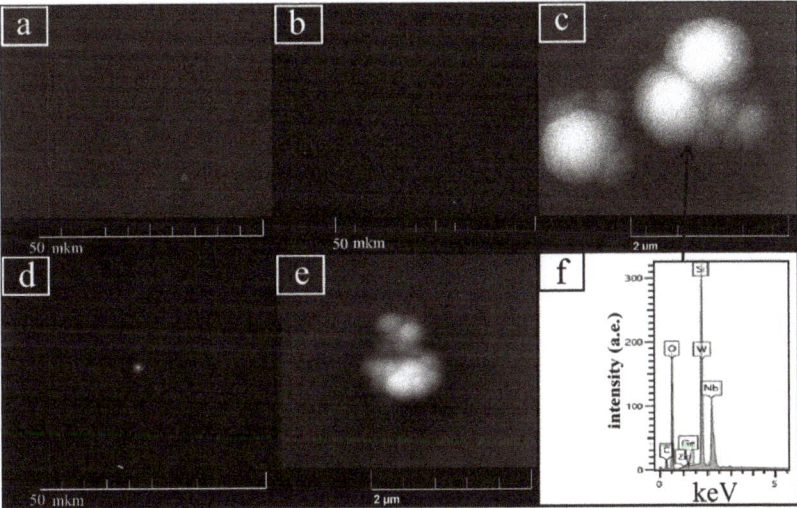

Figure 5. Morphology of interference coatings of samples under study were analyzed using a scanning electron microscope with a Tescan MIRA 3 LMU Schottky cathode equipped with an Oxford Instruments Ultim Max 40 energy dispersive X-ray spectrometer and the chemical composition of the local area of the ZGP_1_SiO$_2$ sample near silicon conglomerates: (**a**) image of the surface of the interference coating of sample ZGP_Al$_2$O$_3$; (**b**) image of the surface of the interference coating of sample ZGP_1_SiO$_2$; (**c**) enlarged image of the silicon conglomerate of the coating sample ZGP_1_SiO$_2$; (**d**) image of the surface of the interference coating of sample ZGP_2_SiO$_2$; (**e**) enlarged image of the silicon conglomerate of the coating sample ZGP_2_SiO$_2$; (**f**) chemical composition of the coating sample ZGP_1_SiO$_2$.

4. Setup Parameters and Technique for Determining the LIDT of the Samples under Study

A Ho:YAG laser generating radiation at a wavelength of 2.097 µm pumped by a cw thulium fiber was the source of radiation. The Ho:YAG laser operated in the active Q-switched mode with a pulse duration τ = 35 ns and a pulse repetition rate of 10 kHz. The measured diameter in all experiments was d = 350 ± 10 µm at the e^{-2} level of the maximum intensity. The maximum average radiation power generated by the Ho:YAG laser was 20 W in a linearly polarized Gaussian beam (parameter $M^2 \leq 1.3$).

The schematic layout of the experimental stand is shown in Figure 6. The power of the incident laser radiation was changed using an attenuator consisting of a half-wave plate ($\lambda/2$) and a polarizing mirror (M1). A Faraday isolator (F.I.) was used to prevent the reflected radiation from entering the laser, which prevented an uncontrolled change in the parameters of the incident radiation. The average laser power (P_{av}) was measured before each experiment with an Ophir power meter (P.M.).

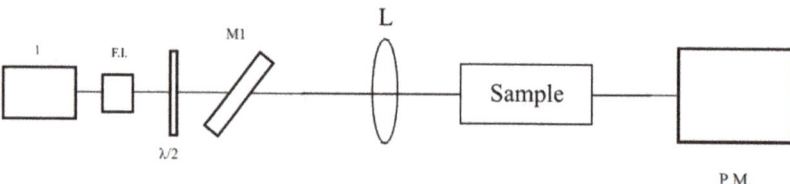

Figure 6. Optical schematic layout of the experimental setup: 1 is the Ho:YAG laser, F.I. is the Faraday isolator, $\lambda/2$ is the half-wave plate, M1 is the polarizing mirror, L is the lens, P.M. is the Ophir power meter.

According to the international standard ISO11146 [23], the energy density of laser radiation was determined by the following expression:

$$W = 8\, P_{av}/(f\pi d^2), \tag{1}$$

where d is the diameter of the laser beam.

The "R-on-1" technique was used to determine the LIDT of the samples, which requires less space on the sample surface compared to the "S-on-1" technique and, therefore, can be used for samples with a limited aperture; however, it is considered coarser [24]. The essence of this technique is that each individual region of the crystal is irradiated with laser radiation with a sequential increase in the intensity of the laser radiation until an optical breakdown occurs or a predetermined value of the energy density is reached. In our work, the study was carried out with an exposure duration τ_{ex} = 5 s. The sample under study was exposed to packets of laser pulses with a fixed energy density level, which did not cause damage to the crystal surface. Then, the energy density level was increased with a step of ~0.1 J/cm². The experiment was terminated when visible damage appeared on one of the surfaces of the nonlinear element. Then, the sample was moved 0.5 mm in height or width using a two-dimensional movement; the experiment was repeated five times. The optical breakdown probability was obtained by plotting the cumulative probability versus the optical breakdown energy density. The value of the LIDT (W_{0d}) was taken to be the energy density corresponding to the approximation of the optical breakdown probability to zero. Figure 7 shows the results of measuring the LIDT using the R-on-1 technique. In the presented plots, the ordinate represents the probability of optical breakdown in relative units, normalized to unity, and the abscissa represents the energy density of the testing laser radiation.

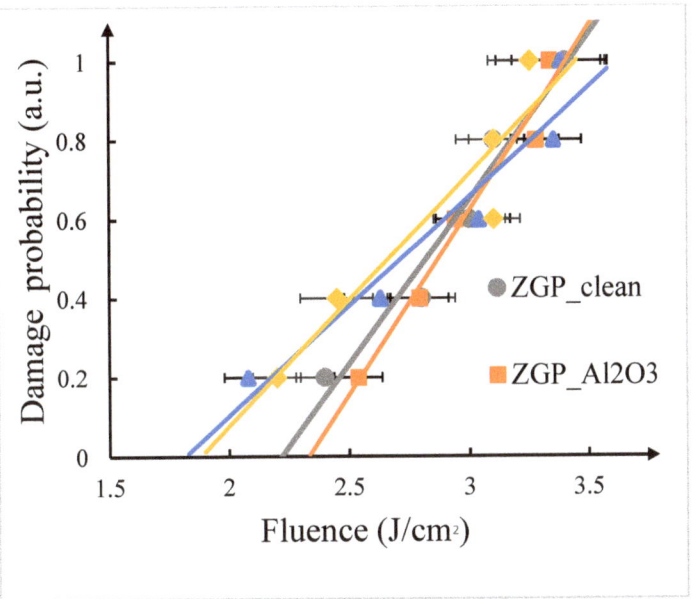

Figure 7. Dependence of the optical breakdown probability of samples ZGP_clean, ZGP_Al2O3 and ZGP_1_SiO2, ZGP_2_SiO2 on the energy density of the incident laser radiation.

5. Experimental Results and Their Discussion

The LIDT values for samples ZGP_clean (without antireflection coatings), ZGP_Al2O3 (with a coating based on alternating layers of Nb_2O_5 and Al_2O_3), ZGP_1_SiO2, and ZGP_2_SiO2 (with a coating based on alternating layers of Nb_2O_5 and SiO_2) (Figure 7) were determined according to the R-on-1 technique, using the experimental stand shown in Figure 6.

Thus, the LIDT for sample ZGP_clean (without antireflection coatings) was $W_{0d} = 2.23$ J/cm^2. The LIDT for sample ZGP_Al2O3 (coated with alternating layers Nb_2O_5 and Al_2O_3) was $W_{0d} = 2.35$ J/cm^2. The LIDT for sample ZGP_1_SiO2 (coated with alternating layers Nb_2O_5 and SiO_2) was $W_{0d} = 1.8$ J/cm^2. The LIDT for sample ZGP_2_SiO2 (coated with alternating layers Nb_2O_5 and SiO_2) was $W_{0d} = 1.86$ J/cm^2. All the samples under study had low absorption at the wavelength of exposure, which indicates a low concentration of point defects that affect the absorption intensity. The presence of binary phosphides and impurity elements and bulk defects was not detected in all the samples under study, which indicates a good quality of the crystal structure of the single crystal.

In previous studies, other researchers have shown that a decrease in the number of layers of the interference coating and the film thickness has a positive effect on the LIDT of the sample surface [21,25]. Moreover, coatings based on SiO_2 showed high LIDT values for various substrate materials compared to coatings based on Al_2O_3 [25]. As can be seen from the results of our experiments, the difference in the thickness of the coatings and in the number of layers in the antireflection coating based on alternating Nb_2O_3/SiO_2 layers is not significant for LIDT. The results in Figure 7 show that a decrease in the coating thickness and a decrease in the number of layers based on the alternation of Nb_2O_3/SiO_2 materials did not lead to a change in LIDT (the antireflection coating of the ZGP_1_SiO2 sample is four times thicker than the ZGP_2_SiO2 sample, and the number of layers is smaller). In our works, we also reduced the rate of the silicon target sputtering. A twofold decrease in the velocity, to 1–1.2 A/s, did not show a significant change in the radiation resistance parameter of the finished coating based on the pair of Nb_2O_3/SiO_2 materials. However, it should be noted that the rate reduction was achieved by changing the anode voltage at the spraying source without changing the gas supply mode. The results obtained on antireflection coatings based on alternating Nb_2O_3/SiO_2 layers can be explained by the lack of oxygen during the deposition of SiO_2 layers, which leads to the incomplete oxidation of the ZGP material deposited on the surface of the samples and the presence of silicon islands in the formed film. This circumstance is most likely the reason for the lower LIDT value compared to the uncoated sample and the absence of the LIDT dependence on the film thickness. This statement is confirmed by the results presented in Figure 5f (according to the chemical formula, SiO_2, the intensity of the spectral lines of the corresponding to O should be higher than the intensity of the spectral lines of silicon (Si), but the opposite picture is actually observed). It is planned to conduct a series of experiments with a change in the gas mode with an increased oxygen content in the plasma to further improve the optical stability of the Nb_2O_3/SiO_2 coating. Moreover, it is planned to make an attempt to sputter a quartz SiO_2 target in an argon plasma with a low oxygen content instead of a pure silicon target, which, according to literature data [26], can give a better characterization of the stoichiometry of the film composition and, as a consequence, a possible increase in LIDT.

Si conglomerates in samples ZGP_1_SiO$_2$ and ZGP_2_SiO$_2$ can be "seed inhomogeneities" for the initialization of the optical breakdown process due to local mechanical stresses and the scattering of laser radiation, which leads to a decrease in the LIDT in comparison with uncoated sample. In turn, the absence of local fluctuations of the composition and mechanical stresses in the case of sample ZGP_Al$_2$O$_3$ leads to an increase in the LIDT in comparison with the uncoated sample due to the "closure" of dangling chemical bonds on the polished surface.

Another factor that can affect the difference between LIDT coatings based on Nb_2O_3/SiO_2 and Nb_2O_3/Al_2O_3 may be the poor adhesion of films based on SiO_2 to the substrate compared to Al_2O_3 [25]. However, one of the adhesion tests that we can carry out is a peel-off test with an adhesive tape (both coatings pass the test—layers do not peel off) and the annealing of samples at a temperature of ~400 °C in an atmosphere for 1 hour, at which both coatings come off the substrates, cracking and flaking. Thus, in this work, we failed to make a correct comparison of the interlayer adhesion and the adhesion of coatings based on Nb_2O_3/SiO_2 and Nb_2O_3/Al_2O_3 to the substrate.

6. Conclusions

Designs of interference coatings based on the alternation of Nb_2O_5 and SiO_2 layers and on the alternation of Nb_2O_5 and Al_2O_3 layers for ZGP single crystals used in parametric light generators generating radiation in the wavelength range of 3–5 µm when pumped by laser radiation at a wavelength of ~2.1 µm have been developed during the research. The technology for sputtering these coatings using the ion-beam sputtering method has been developed. The LIDT at a wavelength of the acting laser radiation of 2.097 µm was determined at a pulse repetition rate of 10 kHz and a pulse duration of 35 ns for ZGP single crystals manufactured by LLC "LOC", which was $W_{0d} = 2.23$ J/cm^2 without antireflection coatings. The LIDT of the sample with a coating based on alternating layers Nb_2O_5 and SiO_2 was $W_{0d} = 1.8$ J/cm^2. The LIDT of the coated sample based on alternating layers Nb_2O_5 and Al_2O_3 was $W_{0d} = 2.35$ J/cm^2.

It has been found that the presence of silicon conglomerates in an interference antireflection coating is a consequence of a lack of oxygen during the deposition of SiO_2 layers, which leads to the incomplete oxidation of the ZGP material deposited on the surface of the samples. It has been found that the presence of silicon conglomerates in an interference antireflection coating leads to a decrease in the LIDT of a nonlinear crystal due to local mechanical stresses and scattering of incident laser radiation, even in the absence of bulk and linear defects of the crystal itself. In turn, the absence of local fluctuations of the composition and mechanical stresses in the case of the sample with a coating based on alternating layers of Nb_2O_5 and Al_2O_3 leads an increase in the LIDT in comparison with an uncoated sample due to the closure of dangling chemical bonds emerging on the polished surface.

The work has been carried out using the equipment of the Tomsk Regional Research Equipment Sharing Center of TSU.

Author Contributions: Conceptualization, N.N.Y., I.K. and M.Z.; methodology, M.Z.; software, V.G. and M.K.; validation, N.A.Y., N.N.Y. and E.S.; formal analysis, M.K.; investigation, E.M., V.G., M.Z. and S.P.; resources, N.N.Y.; data curation, E.S. and E.Z.; writing—original draft preparation, A.P. and E.S.; writing—review and editing, N.N.Y. and S.P.; visualization, M.K.; supervision, N.N.Y.; project administration, N.N.Y.; funding acquisition, N.N.Y. All authors have read and agreed to the published version of the manuscript.

Funding: This study was supported by the Tomsk State University Development Programme («Priority-2030»).

Conflicts of Interest: The authors declare no conflict of interest. The funders had no role in the design of the study; in the collection, analyses, or interpretation of data; in the writing of the manuscript, or in the decision to publish the results.

References

1. Boyd, G.D.; Beuhler, E.; Stortz, F.G. Linear and nonlinear optical properties of $ZnGeP_2$ and CdSe. *Appl. Phys. Lett.* **1971**, *18*, 301–304. [CrossRef]
2. Dmitriev, V.G.; Gurzadyan, G.G.; Nikogosyan, D.N. *Handbook of Nonlinear Optical Crystals*; Springer: New York, NY, USA, 1999; 413p.
3. Beasley, J.D. Thermal conductivities of some novel nonlinear optical materials. *Appl. Opt.* **1994**, *33*, 1000–1003. [CrossRef]
4. Hemming, A.; Richards, J.; Davidson, A.A.; Carmody, N.; Bennetts, S.; Simakov, N.; Haub, J. 99 W mid-IR operation of a ZGP OPO at 25% duty cycle. *Opt. Express* **2013**, *21*, 10062–10069. [CrossRef] [PubMed]
5. Haakestad, M.W.; Fonnum, H.; Lippert, E. Mid-infrared source with 0.2 J pulse energy based on nonlinear conversion of Q-switched pulses in $ZnGeP_2$. *Opt. Express* **2014**, *22*, 8556–8564. [CrossRef]
6. Qian, C.; Yao, B.; Zhao, B.; Liu, G.; Duan, X.; Ju, Y.; Wang, Y. High repetition rate 102 W middle infrared $ZnGeP_2$ master oscillator power amplifier system with thermal lens compensation. *Opt. Lett.* **2019**, *44*, 715–718. [CrossRef] [PubMed]
7. Das, S. Optical parametric oscillator: Status of tunable radiation in mid-IR to IR spectral range based on $ZnGeP_2$ crystal pumped by solid state lasers. *Opt. Quant. Electron.* **2019**, *51*, 70. [CrossRef]
8. Nikogosyan, D. *Nonlinear Optical Crystals: A Complete Survey*; Springer: New York, NY, USA, 2005; 440p.
9. Aggarwal, R.L.; Lax, B. *Nonlinear Infrared Generation*; Shen, Y.R., Ed.; Academic: New York, NY, USA, 1977; 28p.
10. Hildenbrand, A.; Kieleck, C.; Tyazhev, A.; Marchev, G.; Stöppler, G.; Eichhorn, M.; Schunemann, P.G.; Panyutin, V.L.; Petrov, V. Laser damage of the nonlinear crystals $CdSiP_2$ and $ZnGeP_2$ studied with nanosecond pulses at 1064 and 2090 nm. *Opt. Eng.* **2014**, *53*, 122511. [CrossRef]
11. Gribenyukov, A.I.; Dyomin, V.V.; Olshukov, A.S.; Podzyvalov, S.N.; Polovcev, I.G.; Yudin, N.N. Investigation of the process of laserinduced damage of $ZnGeP_2$ crystals using digital holography. *Rus. Phys. J.* **2018**, *61*, 2042–2052. [CrossRef]
12. Chumside, J.H.; Wilson, J.J.; Gribenyukov, A.I.; Shubin, S.F.; Dolgii, S.I.; Andreev, Yu.M.; Zuev, V.V.; Boulder, V. *Co:NOAATechnicalMemorandumERLWPL-224 WPL-224wpl*; IAP: Potsdam, Germany, 1992; p. 18.
13. Andreev Yu, M.; Badikov, V.V.; Voevodin, V.G.; Geiko, L.G.; Geiko, P.P.; Ivashchenko, M.V.; Karapuzikov, A.I.; Sherstov, I.V. Radiation resistance of nonlinear crystals at a wavelength of 9.55 μm. *Quantum Electron.* **2001**, *31*, 1075–1078. [CrossRef]
14. Peterson, R.D.; Schepler, K.L.; Brown, J.L.; Schunemann, P.G. Damage properties of $ZnGeP_2$ at 2 μm. *JOSA B* **1995**, *12*, 2142–2146. [CrossRef]
15. Yudin, N.N.; Antipov, O.L.; Gribenyukov, A.I.; Eranov, I.D.; Podzyvalov, S.N.; Zinoviev, M.M.; Voronin, L.A.; Zhuravleva, E.V.; Zykova, M.P. Effect of postgrowth processing technology and laser radiation parameters at wavelengths of 2091 and 1064 nm on the laser-induced damage threshold in $ZnGeP_2$ single crystal. *Quantum Electron.* **2021**, *51*, 306–316. [CrossRef]
16. Zawilski, K.T.; Setzler, S.D.; Schunemann, P.G.; Pollak, T.M. Increasing the laser-induced damage threshold of single-crystal $ZnGeP_2$. *JOSA B* **2006**, *23*, 2310–2316. [CrossRef]
17. Peng, Y.; Wei, X.; Wang, W. Mid-infrared optical parametric oscillator based on $ZnGeP_2$ pumped by 2-μm laser. *Chin. Opt. Lett.* **2011**, *9*, 061403. [CrossRef]
18. Cheng, X.J.; Zhao, Y.; Qiang, Y.; Zhu, Y.; Guo, L.; Shao, J. Comparison of laser-induced damage in Ta_2O_5 and Nb_2O_5 single-layer films and high. *Chin. Opt. Lett.* **2011**, *9*, 013102. [CrossRef]
19. Zhang, Y.; Xiong, S.; Huang, W. Study on defects in ZnS/YbF_3 infrared coatings on silicon substrates. *Surf. Coat. Technol.* **2017**, *320*, 3–6. [CrossRef]
20. Chen, K.; Hsu, C.; Liu, J.; Liou, Y.; Yang, C. Investigation of Antireflection Nb_2O_5 Thin Films by the Sputtering Method under Different Deposition Parameters. *Micromachines* **2016**, *7*, 151. [CrossRef] [PubMed]

21. Vanyakin, A.V.; Zheleznov, V.I.; Kulevskii, L.A.; Lukashev, A.V.; Morozov, N.P.; Orlov, N.A. Interference optics for lasers and parametric oscillators emitting in the middle-IR range. *Quantum Electron.* **1997**, *24*, 142–144. [CrossRef]
22. Dyomin, V.; Gribenyukov, A.; Davydova, A.; Zinoviev, M.; Olshukov, A.; Podzyvalov, S.; Polovtsev, I.; Yudin, N. Holography of particles for diagnostics tasks [Invited]. *Appl. Opt.* **2019**, *58*, G300–G310. [CrossRef] [PubMed]
23. ISO 11146-1:2005. *Lasers and Laser-Related Equipment—Test Methods for Laser Beam Widths, Divergence Angles and Beam Propagation Ratios*; American National Standards Institute (ANSI): Washington, DC, USA, 2005.
24. "The R-on-1 Test", Lidaris LIDT Service. 2019. Available online: http://lidaris.com/laserdamage-testing/ (accessed on 16 November 2020).
25. Gallais, L.; Commandré, M. Laser-induced damage thresholds of bulk and coatingoptical materials at 1030 nm, 500 fs. *Appl. Opt.* **2013**, *53*, A186–A196. [CrossRef] [PubMed]
26. Telesh, E.; Kasinsky, N.; Tomal, V. Coatings formation by ion beam sputtering of dielectric targets. *Bull. Polotsk State Univ. Ser. C Fundam. Sci.* **2012**, *4*, 23–29.

Article

β-Ga$_2$O$_3$ Used as a Saturable Sbsorber to Realize Passively Q-Switched Laser Output

Baizhong Li [1,2], Qiudi Chen [3], Peixiong Zhang [3,*], Ruifeng Tian [1,2], Lu Zhang [1,2], Qinglin Sai [1], Bin Wang [1], Mingyan Pan [1], Youchen Liu [1], Changtai Xia [1], Zhenqiang Chen [3] and Hongji Qi [1,4,*]

[1] Key Laboratory of Materials for High Power Laser, Shanghai Institute of Optics and Fine Mechanics, Chinese Academy of Sciences, Shanghai 201800, China; lbz446@siom.ac.cn (B.L.); ruifengtian@siom.ac.cn (R.T.); zhanglu@siom.ac.cn (L.Z.); saiql@siom.ac.cn (Q.S.); wangbinmars@siom.ac.cn (B.W.); pmy@siom.ac.cn (M.P.); lyc@siom.ac.cn (Y.L.); xia_ct@siom.ac.cn (C.X.)
[2] Center of Materials Science and Optoelectronics Engineering, University of Chinese Academy of Sciences, Beijing 100049, China
[3] Department of Optoelectronic Engineering, Jinan University, Guangzhou 510632, China; Cqd596918045@163.com (Q.C.); tzqchen@jnu.edu.cn (Z.C.)
[4] Hangzhou Institute of Optics and Fine Mechanics, Hangzhou 311421, China
* Correspondence: pxzhang@jnu.edu.cn (P.Z.); qhj@siom.ac.cn (H.Q.)

Abstract: β-Ga$_2$O$_3$ crystals have attracted great attention in the fields of photonics and photoelectronics because of their ultrawide band gap and high thermal conductivity. Here, a pure β-Ga$_2$O$_3$ crystal was successfully grown by the optical floating zone (OFZ) method, and was used as a saturable absorber to realize a passively Q-switched all-solid-state 1 μm laser for the first time. By placing the as-grown β-Ga$_2$O$_3$ crystal into the resonator of the Nd:GYAP solid-state laser, Q-switched pulses at the center wavelength of 1080.4 nm are generated under a output coupling of 10%. The maximum output power is 191.5 mW, while the shortest pulse width is 606.54 ns, and the maximum repetition frequency is 344.06 kHz. The maximum pulse energy and peak power are 0.567 μJ and 0.93 W, respectively. Our experimental results show that the β-Ga$_2$O$_3$ crystal has great potential in the development of an all-solid-state 1 μm pulsed laser.

Keywords: β-Ga$_2$O$_3$ crystal; optical floating zone; saturable absorber; Q-switch

1. Introduction

It is well known that saturable absorbers play an important role in Q-switching and mode locking operation [1–4]. Therefore, the development of different kinds of saturable absorbers as passive Q-switching devices, to achieve high-quality pulsed laser output, has always been a hot research field. At present, the research on saturable absorbers is in full swing. There are not only traditional saturable absorbers, such as dyes and transition metal ion-doped crystals, but also some new phase change materials, including bulk semiconductors and two-dimensional materials [5–9]. The pulsed laser realized by some of the materials has important application prospects in industrial processing, high-energy lasers, scientific research, and so on [10–12]. Particularly, ~1 μm near-infrared lasers, which have the advantages of high pulse energy and high peak power, can be widely used in space communication, nonlinear spectroscopy, biomedicine, military, and many other fields [13–15]. However, traditional materials often have their own shortcomings, such as limited types, single wavelength, and long-term operation stability, which need to be improved. Therefore, how to develop a stable, reliable and efficient new saturable absorber for application in the ~1 μm near-infrared band is a problem worthy of further discussion.

Ga$_2$O$_3$ is a semiconductor material with an ultra-wide band gap (~4.8 eV) and high conductivity [16,17]. Therefore, Ga$_2$O$_3$ is an electronic and optical material with great potential. Because of its unique physical and chemical properties, it has received great attention from researchers in different areas, so it has been applied in many fields, including in photo-detectors, photo-catalysis, field effect transistors, and so on [18–21]. Ga_2O_3 polymorphic, similar to Al_2O_3, which makes it particularly interesting in applications. β-Ga_2O_3 has a monoclinic structure and is the most stable phase, both physically and chemically [22,23]. β-Ga_2O_3 also inherits the excellent physical and chemical properties common to all phases of Ga_2O_3 [24,25]. These excellent properties make it clear that β-Ga_2O_3 has great potential for application in saturable absorbers. However, as far as we know, there are still no reports on the application of pure β-Ga_2O_3 crystals to saturable absorbers, and even the application of other oxide materials to saturable absorbers is rarely reported.

In terms of the β-Ga_2O_3 crystal growth method, the common large-size crystal growth method is a melting method, similarly to the Czochralski method, EFG method, Bridgman method, and so on [26–28]. We have successfully grown high-quality β-Ga_2O_3 using the optical floating zone (OFZ) method, which is a method of growing crystals without a crucible, and is usually used to study and explore the properties of materials. This method has the advantages of simple operation, few equipment requirements, and the ability to even grow crystals in the air environment. Compared with the Czochralski method, the floating zone method has the advantages of simple operation and a short cycle, and can effectively reduce the economic cost of crystal growth. Compared with the Verneuil process, the crystal quality is better [29,30]. This method solves many technical problems, such as complex equipment, difficult operation, easy introduction of impurities, inability to guarantee the growth quality, and so on. We systematically characterized the chemical and optical properties of the synthesized β-Ga_2O_3. The crystal is of good quality; it is pure and crack free. At the same time, we realized the optical modulation of β-Ga_2O_3 in pulse laser generation for the first time on the laser device with b-cut Nd:GYAP (Nd:$Gd_{0.1}Y_{0.9}AlO_3$) as the laser medium [31]. The maximum average output power is 195.1 mW, which is obtained at 1080.4 nm. The corresponding shortest pulse duration is 606.54 ns and the maximum pulse repetition rate is 344.06 kHz. The maximum single pulse energy is 0.567 µJ and the maximum peak power is 0.93 W. From the experimental results, we have obtained a relatively stable pulsed laser with a short pulse width and large repetition frequency, which shows that β-Ga_2O_3 has good saturable absorption properties. Compared with the common two-dimensional material saturable absorbers of around 1 µm, such as graphene, graphene oxide, black phosphorus (BP), topological insulators (TI), and transition metal dichalcogenides (TMDs) [5,32,33], β-Ga_2O_3, as a crystal plate, is easy to mass produce, the product performance is stable, and it is not easy to damage. The experimental results are also easy to replicate. At the same time, compared with other crystal planes used as saturable absorbers of around 1 µm, such as Cr^{4+}:YAG, Cr^{2+}:ZnS, Co^{2+}:$LaMgAl_{11}O_{19}$, and V^{3+}:YAG, the thermal conductivity of the β-Ga_2O_3 crystal is about 27 $W \cdot m^{-1} \cdot K^{-1}$, which is much larger than that of ZnS(0.561 $W \cdot m^{-1} \cdot K^{-1}$), $LaMgAl_{11}O_{19}$(2.55 $W \cdot m^{-1} \cdot K^{-1}$), and YAG(12.9 $W \cdot m^{-1} \cdot K^{-1}$) [34–40]. This indicates that the β-Ga_2O_3 crystal is favorable for the output of laser pulses with a high peak power and high repetition rate. We believe that our work will provide an important reference for the potential applications of nonlinear optical devices related to crystal growth and optical modulation.

2. The Preparation and Characterization of β-Ga_2O_3

The β-Ga_2O_3 single crystal was grown by the optical floating zone (OFZ) method, using a Quantum Design IRF01-001-00 infrared image furnace (IR Image Furnace G3, Quantum Design Japan). Ga_2O_3 powder (purity: 99.9999%, Alfa Aesar) was employed as the raw material. The raw material was pressed into a rod using a cold isostatic press. The rod was subsequently sintered at 1400 °C for 10 h in air. Moreover, a <010> oriented crystal was used as the seed. Growth was carried out using a Quantum Design IRF01-001-00 infrared image furnace. The sintered rod and seed were rotated at 10 rpm in opposite directions, and the crystal was grown in flowing air at a speed of 6 mm/h. Figure 1 shows a photo of the as-grown β-Ga_2O_3 single crystal. After growth, the as-grown sample was cut into 6 × 5 × 1 mm³ wafers and subjected to chemical mechanical polishing to form a 0.5-millimeter-thick wafer parallel to the (100) plane to measure its optical properties.

Figure 1. The photo of as-grown β-Ga$_2$O$_3$ single crystal.

The X-ray rocking curve was measured using a Bruker D8 Discover X-ray diffractometer with a Cu Kα line at 40 kV and 40 mA. The optical transmittance spectrum was collected using a Lambda 1050+ UV/Vis/NIR spectrometer (PerkinElmer). Figure 2 shows the X-ray rocking curve of the β-Ga$_2$O$_3$ (400) plane. The full width at half maximum (FWHM) is 100.8 arcsec. This shows that the β-Ga$_2$O$_3$ crystal is a single crystal with good crystallization quality. Figure 3 shows the optical transmission spectrum of the β-Ga$_2$O$_3$ single crystal. The β-Ga$_2$O$_3$ single crystal wafer indicates high transmittance, between 80% and 82%, from the visible wavelength to the infrared (IR) wavelength region. The transmittance spectrum exhibits a cutoff absorption edge at around 255 nm. This was a result of the intrinsic absorption caused by the transition from the valence band to the conduction band [41].

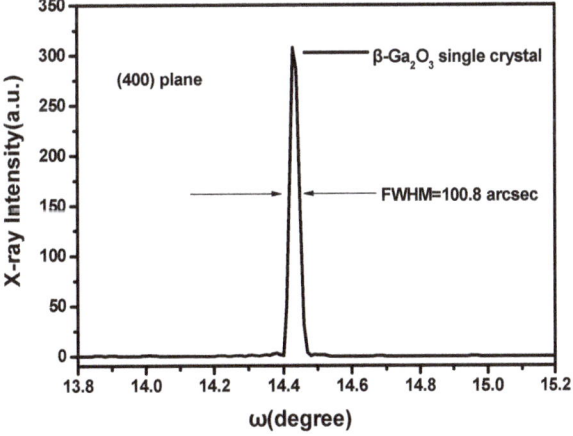

Figure 2. X-ray rocking curve of the β-Ga$_2$O$_3$ single crystal ((400) plane).

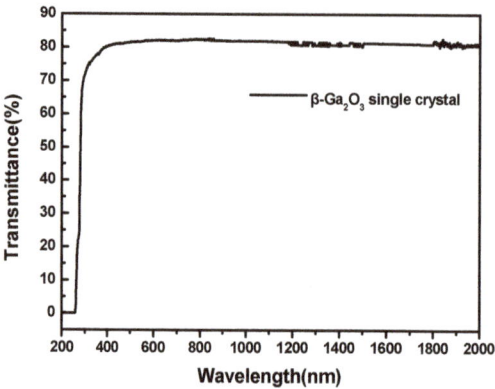

Figure 3. The optical transmission spectrum of the β-Ga$_2$O$_3$ single crystal.

3. Experimental Results and Discussion

To study the saturable absorption characteristics of β-Ga$_2$O$_3$ SA in the 1 μm wavelength region, a passively Q-switched laser, composed of a Nd:GYAP laser crystal and β-Ga$_2$O$_3$ SA, was constructed, as shown in Figure 4. The crystal was cut to a 4 × 4 × 5 mm^3 cuboid, along the b axis. The pump source is an 808 nm fiber-coupled semiconductor laser diode (LD), and the core diameter is 400 μm, with an aperture of 0.22. Using an optical imaging system (1:1 imaging module), the spot radius of the pump laser beam focused on the Nd:GYAP crystal is 200 μm. The resonator uses an input mirror M1 with high reflection from 1050 nm to 1100 nm and high transmittance from 800 nm to 820 nm, and an output mirror M2 with 10% transmittance from 1050 nm to 1100 nm. β-Ga$_2$O$_3$ SA is inserted between M2 and the gain crystal. During normal operation, the Nd:GYAP crystal is wrapped in indium foil and maintained at 17 °C by a chiller to minimize the thermal lens effect.

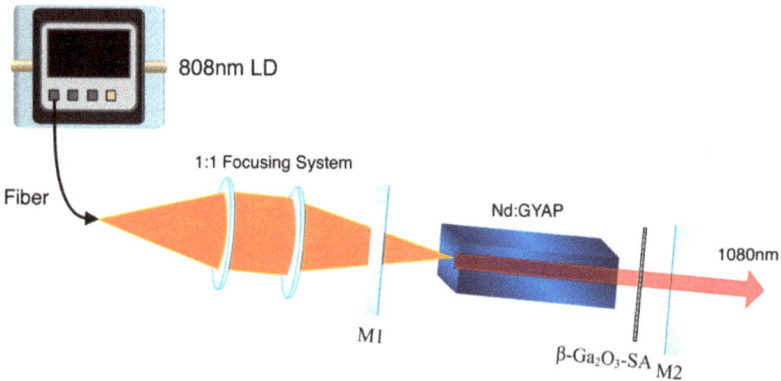

Figure 4. Schematic experimental setup of the β-Ga$_2$O$_3$ SA Q-switched Nd:GYAP laser.

When no saturable absorber is added, a continuous wave (CW), with a threshold of 0.759 W, is obtained. Then, we insert the prepared β-Ga$_2$O$_3$ SA into the laser cavity to realize the Q-switched pulse. During the experiment, the average output power of the CW and Q-switched lasers is measured as a function of pump power. It is obvious from Figure 5 that the average output power of the two groups increases linearly with the pump power. After linear fitting, the slope efficiency of the CW and Q-switched lasers are 28.6% and 10.2%, respectively. When the absorption pump power of the CW laser is 3.75 W, the maximum output power is 0.9262 W, and, at this time, it reaches the highest Q-switched average output power of 195.1 mW. The slope efficiency of the Q-switched laser is lower than that of the continuous laser, which is mainly due to the unsaturated absorption loss of Ga_2O_3 SA.

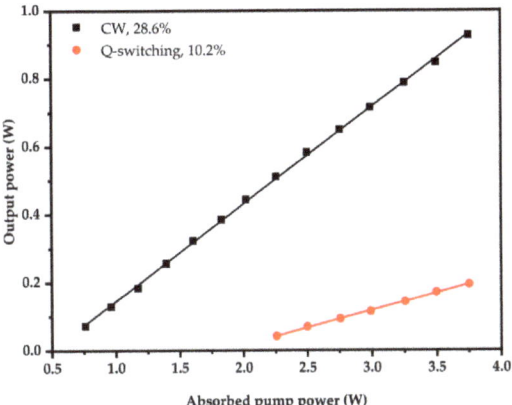

Figure 5. The average output power of Nd:GYAP laser in different operation regimes.

The pulse generation is due to the sudden decrease in the laser oscillation threshold, caused by the complete saturated absorption of SA. After the first pulse output, the absorption of SA returns to a higher initial value, and the inverted particle swarm can accumulate again to prepare for the formation of the next pulse. According to the principle of passive Q-switched pulses, at a high pump power, a short saturation absorption period and strong stimulated radiation are helpful to produce pulses with a high repetition rate and narrow pulse width, respectively. Therefore, with the increase in pump power, the duration of the Q-switched pulse narrows, and the number of pulses in the same time period increases. Figure 6 shows the variation in pulse width and repetition frequency with increasing pump power. With the increase in pump power from 2.256 W to 3.751 W, the repetition rate curve shows a continuous upward trend from 88.67 kHz to 344.06 kHz, while the width of a single pulse decreases from 2030.39 ns to 606.54 ns. Figure 7 shows the oscilloscope image at the highest repetition frequency and the shortest pulse. Through the relatively neat pulse sequence in the picture, we also know that we have obtained a relatively stable and neat pulse laser.

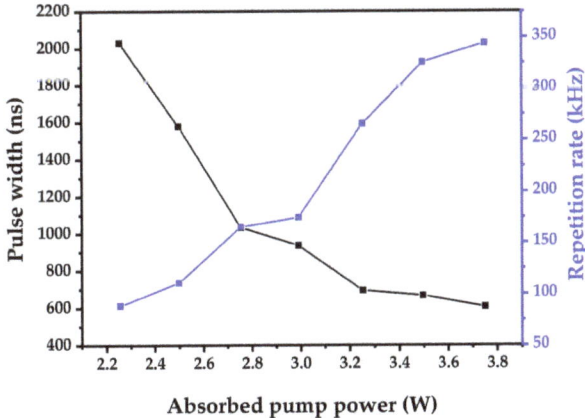

Figure 6. Dependences of pulse width and repetition rate on absorbed pump power.

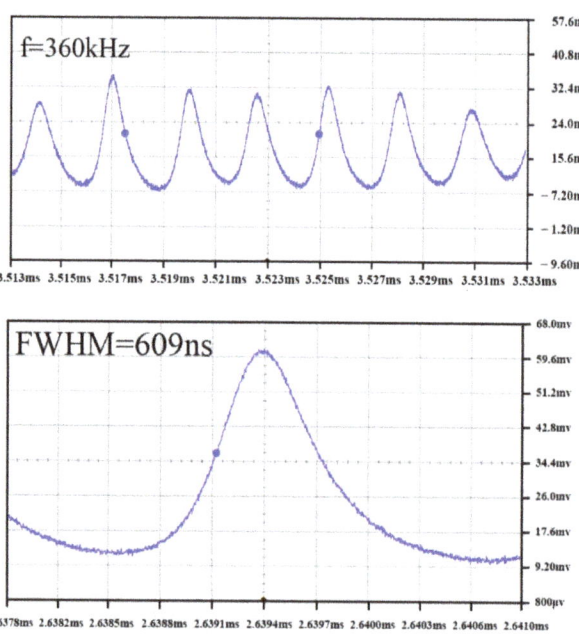

Figure 7. Temporal pulse train and single pulse profile from the β-Ga$_2$O$_3$-SA Q-switched Nd:GYAP laser.

Based on the average output power, pulse width, and repetition frequency, the corresponding single pulse energy and peak power of the Q-switched laser are calculated. The single pulse energy and peak power increase with the increase in pump power, which proves that our laser output is in Q-switched mode rather than relaxation oscillation mode. When the pump power is 3.75 W, the maximum single pulse energy is 0.567 µJ and the maximum peak power is 0.93 W. We also measured the central wavelength of laser emission. Figure 8 shows the emission wavelength in the Q-switched region, with a peak at about 1080.4 nm. The inserted β-Ga$_2$O$_3$ SA does not change the emission wavelength of the Nd:GYAP laser. Furthermore, as a stable bulk sample, as long as the laser output power and temperature do not reach the threshold of damage or crack, the β-Ga$_2$O$_3$ saturable absorber can continuously and stably output the pulsed laser and can be reused many times.

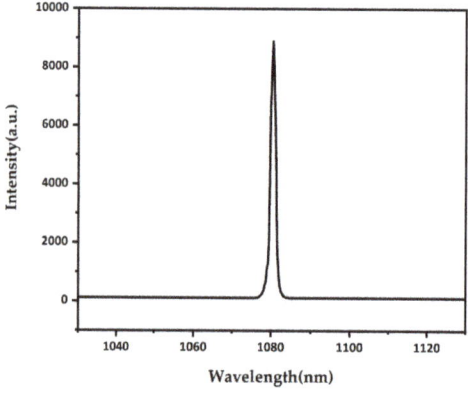

Figure 8. The laser emission spectrum of Nd:GYAP laser in Q-switching regimes.

4. Conclusions

In summary, we have successfully grown high-quality β-Ga₂O₃ used the OFZ method, and have used them as saturable absorbers, to realize the output of the pulsed laser, for the first time. The synthetic method is simple and practical, with low cost and low environmental requirements, and the grown crystal is pure and crack free. At the same time, the β-Ga₂O₃ crystal is applied to a Nd:GYAP solid-state laser for the first time, and the pulsed laser output is realized. The maximum average output power of 195.1 mW is obtained at 1080.4 nm. The corresponding minimum pulse width is 606.54 ns and the maximum pulse repetition frequency is 344.06 kHz. Our results will promote the research of more Q-switched crystals and expand their potential applications in the field of

Author Contributions: Conceptualization, B.L., P.Z. and H.Q.; data curation, B.L. and Q.C.; formal analysis, B.L., Q.C. and P.Z.; funding acquisition, P.Z., Q.S., B.W., M.P., C.X., Z.C. and H.Q.; investigation, B.L., R.T., L.Z., Q.C. and P.Z.; project administration, Y.L.; resources, Q.S., B.W., M.P. and C.X.; supervision, Z.C. and H.Q.; writing-original draft, B.L. and Q.C.; writing-review and editing, B.L., P.Z. and H.Q. All authors have read and agreed to the published version of the manuscript.

Funding: This work was supported by the National Natural Science Foundation of China (NSFC) (52072183, 52002386, 51972319, 51972149, 51872307, 61935010); the Shanghai Science and Technology Commission (20511107400); the Fundamental Research Funds for the Central Universities (21620445).

Institutional Review Board Statement: Not applicable.

Data Availability Statement: Data available in a publicly accessible repository.

Acknowledgments: In particular, we would like to thank Hangzhou Fujia Gallium Technology Co., Ltd. for its help in crystal growth and processing.

Conflicts of Interest: The authors declare no conflict of interest.

References

1. Namour, M.; Mobadder, M.; Magnin, D.; Peremans, A.; Verspecht, T.; Teughels, W.; Lamard, L.; Nammour, S.; Rompen, E. Q-Switch Nd:YAG laser-assisted decontamination of implant surface. *Dent. J.* **2019**, *7*, 99. [CrossRef]
2. Namour, M.; Verspecht, T.; Mobadder, M.; Teughels, W.; Peremans, A.; Nammour, S.; Rompen, E. Q-Switch Nd:YAG laser-assisted elimination of multi-species biofilm on titanium surfaces. *Materials* **2020**, *13*, 1573. [CrossRef] [PubMed]
3. Cong, Z.; Liu, Z.; Qin, Z.; Zhang, X.; Wang, S.; Rao, H.; Fu, Q. RTP Q-switched single-longitudinal-mode Nd:YAG laser with a twisted-mode cavity. *Appl. Opt.* **2015**, *54*, 5143–5146. [CrossRef] [PubMed]
4. Suzuki, M.; Boyraz, O.; Asghari, H.; Jalali, B. Spectral dynamics on saturable absorber in mode-locking with time stretch spectroscopy. *Sci. Rep.* **2020**, *10*, 14360. [CrossRef]
5. Ren, C.; Deng, X.; Hu, W.; Li, J.; Miao, X.; Xiao, S.; Liu, H.; Fan, Q.; Wang, K.; He, T. A near-infrared I emissive dye: Toward the application of saturable absorber and multiphoton fluorescence microscopy at the deep-tissue imaging window. *Chem. Commun. (Camb.)* **2019**, *25*, 5111–5114. [CrossRef] [PubMed]
6. Gerislioglu, B.; Bakan, G.; Ahuja, R.; Adam, J.; Mishra, Y.K.; Ahmadivand, A. The role of Ge₂Sb₂Te₅ in enhancing the performance of functional plasmonic devices. *Mater. Today Phys.* **2020**, *12*, 100178. [CrossRef]
7. Wang, M.; Zheng, Y.; Guo, L.; Chen, X.; Zhang, H.; Li, D. Nonlinear optical properties of zirconium diselenide and its ultra-fast modulator application. *Nanomaterials* **2019**, *9*, 1419. [CrossRef]
8. Yang, J.; Tian, K.; Li, Y.; Dou, X.; Ma, Y.; Han, W.; Xu, H.; Liu, J. Few-layer Bi₂Te₃: An effective 2D saturable absorber for passive Q-switching of compact solid-state lasers in the 1-μm region. *Opt. Express* **2018**, *26*, 21379–21389. [CrossRef]
9. Ge, W.; Zhang, H.; Wang, J.; Cheng, X.; Jiang, M.; Du, C.; Yuan, S. Pulsed laser output of LD-end-pumped 1.34 mum Nd:GdVO₄ laser with Co:LaMgAl₁₁O₁₉ crystal as saturable absorber. *Opt. Express* **2005**, *13*, 3883–3889. [CrossRef]
10. Malinauskas, M.; Žukauskas, A.; Hasegawa, S.; Hayasaki, Y.; Mizeikis, V.; Buividas, R.; Juodkazis, S. Ultrafast laser processing of materials: From science to industry. *Light Sci. Appl.* **2016**, *5*, 16133. [CrossRef]
11. Kouno, A.; Watanabe, S.; Hongo, T.; Yao, K.; Satake, K.; Okiji, T. Effect of pulse energy, pulse frequency, and tip diameter on intra-canal vaporized bubble kinetics and apical pressure during laser-activated irrigation using Er:YAG Laser. *Photomed. Laser Surg.* **2020**, *38*, 431–437. [CrossRef]
12. Ge, Z.; Saito, T.; Kurose, M.; Kanda, H.; Arakawa, K.; Takeda, M. Precision interferometry for measuring wavefronts of multi-wavelength optical pickups. *Opt. Express* **2008**, *16*, 133–143. [CrossRef]
13. Griffith, R.; Simmons, B.; Bray, F.; Falto-Aizpurua, L.; Abyaneh, M.; Nouri, K. 1064 nm Q-switched Nd:YAG laser for the treatment of Argyria: A systematic review. *J. Eur. Acad. Dermatol. Venereol.* **2015**, *29*, 2100–2103. [CrossRef]

14. Chang, Y.; Lee, J.; Jhon, Y.; Lee, J. Active Q-switching in an erbium-doped fiber laser using an ultrafast silicon-based variable optical attenuator. *Opt. Express* **2011**, *19*, 26911–26916. [CrossRef] [PubMed]
15. Cabalín, L.; González, A.; Lazic, V.; Laserna, J. Deep ablation and depth profiling by laser-induced breakdown spectroscopy (LIBS) employing multi-pulse laser excitation: Application to galvanized steel. *Appl. Spectrosc.* **2011**, *65*, 797–805. [CrossRef] [PubMed]
16. Kim, S.; Lee, H.; Oh, S.; Noh, B.; Park, S.; Im, Y.; Son, S.; Song, Y.; Kim, K. Transparent conductive electrodes of β-Ga_2O_3/Ag/β-Ga_2O_3 multilayer for ultraviolet emitters. *J. Nanosci. Nanotechnol.* **2019**, *19*, 6328–6333. [CrossRef]
17. An, Y.; Shen, X.; Hao, Y.; Guo, P.; Tang, W. Enhanced resistance switching of Ga_2O_3 thin films by ultraviolet radiation. *J. Nanosci. Nanotechnol.* **2020**, *20*, 3283–3286. [CrossRef] [PubMed]
18. Ma, J.; Yoo, G. Electrical properties of top-gate β-Ga_2O_3 nanomembrane metal-semiconductor field-effect transistor. *J. Nanosci. Nanotechnol.* **2020**, *20*, 516–519. [CrossRef]
19. Bae, H.; Yoo, T.; Yoon, Y.; Lee, I.; Kim, J.; Cho, B.; Hwang, W. High-aspect ratio β-Ga_2O_3 nanorods via hydrothermal synthesis. *Nanomaterials* **2018**, *8*, 594. [CrossRef]
20. Long, X.; Niu, W.; Wan, L.; Chen, X.; Cui, H.; Sai, Q.; Xia, C.; Devki, N.T.; Feng, Z. Optical and Electronic Energy Band Properties of Nb-Doped β-Ga_2O_3 crystals. *Crystals* **2021**, *11*, 135. [CrossRef]
21. Hisatomi, T.; Brillet, J.; Cornuz, M.; Le Formal, F.; Tétreault, N.; Sivula, K.; Grätzel, M.A. Ga_2O_3 underlayer as an isomorphic template for ultrathin hematite films toward efficient photoelectrochemical water splitting. *Faraday Discuss.* **2012**, *155*, 223–232. [CrossRef]
22. Zhou, H.; Zeng, S.; Zhang, J.; Liu, Z.; Feng, Q.; Xu, S.; Zhang, J.; Hao, Y. Comprehensive Study and Optimization of Implementing p-NiO in β-Ga_2O_3 Based Diodes via TCAD Simulation. *Crystals* **2021**, *11*, 1186. [CrossRef]
23. Reddy, L.; Ko, Y.; Yu, J. Hydrothermal synthesis and photocatalytic property of β-Ga_2O_3 nanorods. *Nanoscale Res. Lett.* **2015**, *10*, 364. [CrossRef]
24. Huan, Y.; Sun, S.; Gu, C.; Liu, W.; Ding, S.; Yu, H.; Xia, C.; Zhang, D. Recent advances in β-Ga_2O_3-Metal contacts. *Nanoscale Res. Lett.* **2018**, *13*, 246. [CrossRef]
25. Cui, W.; Ren, Q.; Zhi, Y.; Zhao, X.; Wu, Z.; Li, P.; Tang, W. Optimization of growth temperature of β-Ga_2O_3 thin films for solar-blind photodetectors. *J. Nanosci. Nanotechnol.* **2018**, *18*, 3613–3618. [CrossRef]
26. Yeom, T.; Lim, A. Study of nuclear quadrupole interactions and quadrupole Raman processes of ^{69}Ga and ^{71}Ga in a β-Ga_2O_3:Cr^{3+} single crystal. *J. Magn. Reson.* **2009**, *200*, 261–266. [CrossRef]
27. Xue, H.; He, Q.; Jian, G.; Long, S.; Pang, T.; Liu, M. An overview of the ultrawide bandgap Ga_2O_3 Semiconductor-Based schottky barrier diode for power electronics application. *Nanoscale Res. Lett.* **2018**, *13*, 290. [CrossRef]
28. Hoshikawa, K.; Kobayashi, T.; Ohba, E.; Kobayashi, T. 50mm diameter Sn-doped (001) β-Ga_2O_3 crystal growth using the Vertical Bridegman Technique in ambient air. *J. Cryst. Growth* **2020**, *546*, 125778. [CrossRef]
29. Abbene, L.; Principato, F.; Gerardi, G.; Buttacavoli, A.; Cascio, D.; Bettelli, M.; Amadè, N.; Seller, P.; Veale, M.; Fox, O.; et al. Room-temperature X-ray response of cadmium-zinc-telluride pixel detectors grown by the vertical Bridgman technique. *J. Synchrotron Radiat.* **2020**, *27*, 319–328. [CrossRef] [PubMed]
30. Kozhemyakin, G.; Nemets, L.; Bulankina, A. Simulation of ultrasound influence on melt convection for the growth of $Ga_xIn_{1-x}Sb$ and Si single crystals by the Czochralski method. *Ultrasonics* **2014**, *54*, 2165–2168. [CrossRef] [PubMed]
31. Zhou, H.; Zhu, S.; Li, Z.; Yin, H.; Zhang, P.; Chen, Z.; Fu, S.; Zhang, Q.; Lv, Q. Investigation on 1.0 and 1.3 µm laser performance of Nd^{3+}: GYAP crystal. *Opt. Laser Technol.* **2019**, *119*, 105601. [CrossRef]
32. Sun, X.; Nie, H.; He, J.; Zhao, R.; Su, X.; Wang, Y.; Zhang, B.; Wang, R.; Yang, K. Passively mode-locked 1.34 µm bulk laser based on few-layer black phosphorus saturable absorber. *Opt. Express* **2017**, *25*, 20025–20032. [CrossRef]
33. Fan, M.; Li, T.; Zhao, S.; Li, G.; Ma, H.; Gao, X.; Kränkel, C.; Huber, G. Watt-level passively Q-switched Er:Lu_2O_3 laser at 2.84 m using MoS_2. *Opt. Lett.* **2016**, *41*, 540–543. [CrossRef]
34. Zhu, H.; Chen, Y.; Lin, Y.; Gong, X.; Luo, Z.; Huang, Y. Efficient quasi-continuous-wave and passively Q-switched laser operation of a Nd^{3+}:$BaGd_2(MoO_4)_4$ cleavage plate. *Appl. Opt.* **2008**, *47*, 531–535. [CrossRef] [PubMed]
35. Yao, B.; Yuan, J.; Li, J.; Dai, T.; Duan, X.; Shen, Y.; Cui, Z.; Pan, Y. High-power Cr^{2+}:ZnS saturable absorber passively Q-switched Ho:YAG ceramic laser and its application to pumping of a mid-IR OPO. *Opt. Lett.* **2015**, *40*, 348–351. [CrossRef]
36. Li, P.; Li, Y.; Sun, Y.; Hou, X.; Zhang, H.; Wang, J. Passively Q-switched 1.34 mum Nd:$Y_xGd_{(1-x)}VO_{(4)}$ laser with Co^{2+}:$LaMgAl_{(11)}O_{(19)}$ saturable absorber. *Opt. Express* **2006**, *14*, 7730–7736. [CrossRef] [PubMed]
37. Huang, H.T.; He, J.L.; Zhang, B.T.; Yang, J.F.; Xu, J.L.; Zuo, C.H.; Tao, X.T. V^{3+}:YAG as the saturable absorber for a diode-pumped quasi-three-level dual-wavelength Nd:GGG laser. *Opt. Express* **2010**, *18*, 3352–3357. [CrossRef]
38. Kharazmi, A.; Faraji, N.; Mat Hussin, R.; Saion, E.; Yunus, W.M.; Behzad, K. Structural, optical, opto-thermal and thermal properties of ZnS-PVA nanofluids synthesized through a radiolytic approach. *Beilstein J. Nanotechnol.* **2015**, *6*, 529–536. [CrossRef]
39. Guo, Z.; Verma, A.; Wu, X.F.; Sun, F.Y.; Hickman, A.; Masui, T.; Kuramata, A.; Higashiwaki, M.; Jena, D.; Luo, T.F. Anisotropic thermal conductivity in single crystal β-gallium oxide. *Appl. Phys. Lett.* **2015**, *106*, 1–5. [CrossRef]
40. Sun, J.B.; Niu, L.F.; Hui, Y.L.; Chen, W.Y.; Wang, X.L.; Lu, J.S.; Wei, H. Thermal conductivity and compatibility of $LaMgAl_{11}O_{19}$/$LaPO_4$ composites. *Ceram. Int.* **2020**, *46*, 27967–27972. [CrossRef]
41. Li, P.; Bu, Y.; Chen, D.; Sai, Q.; Qi, H. Investigation of the crack extending downward along the seed of the β-Ga_2O_3 crystal grown by the EFG method. *CrystEngComm* **2021**, *23*, 6300–6306. [CrossRef]

Article

Energy Transfer and Cross-Relaxation Induced Efficient 2.78 µm Emission in Er^{3+}/Tm^{3+}: PbF_2 mid-Infrared Laser Crystal

Jiayu Liao [1,2,3,†], Qiudi Chen [1,2,3,†], Xiaochen Niu [1,2,3], Peixiong Zhang [1,2,3,*], Huiyu Tan [1,2,3], Fengkai Ma [1,2,3], Zhen Li [1,2,3], Siqi Zhu [1,2,3], Yin Hang [4], Qiguo Yang [5] and Zhenqiang Chen [1,2,3]

1. Guangdong Provincial Key Laboratory of Optical Fiber Sensing and Communications, Guangzhou 510630, China; jyliao@stu2019.jnu.edu.cn (J.L.); Cqd596918045@163.com (Q.C.); nxc_n1u@163.com (X.N.); thy@stu2020.jnu.edu.cn (H.T.); hkai80@163.com (F.M.); ailz268@126.com (Z.L.); tzhusiqi@jnu.edu.cn (S.Z.); tzqchen@jnu.edu.cn (Z.C.)
2. Guangdong Provincial Engineering Research Center of Crystal and Laser Technology, Guangzhou 510632, China
3. Department of Optoelectronic Engineering, Jinan University, Guangzhou 510632, China
4. Key Laboratory of High Power Laser Materials, Shanghai Institute of Optics and Fine Mechanics, Chinese Academy of Sciences, Shanghai 201800, China; yhang@siom.ac.cn
5. Guangdong Provincial Key Laboratory of Industrial Ultrashort Pulse Laser Technology, Shenzhen 518055, China; yangqiguo@126.com
* Correspondence: pxzhang@jnu.edu.cn
† These authors contribute equally to this work.

Citation: Liao, J.; Chen, Q.; Niu, X.; Zhang, P.; Tan, H.; Ma, F.; Li, Z.; Zhu, S.; Hang, Y.; Yang, Q.; et al. Energy Transfer and Cross-Relaxation Induced Efficient 2.78 µm Emission in Er^{3+}/Tm^{3+}: PbF_2 mid-Infrared Laser Crystal. Crystals 2021, 11, 1024. https://doi.org/10.3390/cryst 11091024

Academic Editors: Ludmila Isaenko and Anna Paola Caricato

Received: 22 July 2021
Accepted: 24 August 2021
Published: 26 August 2021

Publisher's Note: MDPI stays neutral with regard to jurisdictional claims in published maps and institutional affiliations.

Copyright: © 2021 by the authors. Licensee MDPI, Basel, Switzerland. This article is an open access article distributed under the terms and conditions of the Creative Commons Attribution (CC BY) license (https://creativecommons.org/licenses/by/4.0/).

Abstract: An efficient enhancement of 2.78 µm emission from the transition of Er^{3+}: $^4I_{11/2} \to {}^4I_{13/2}$ by Tm^{3+} introduction in the Er/Tm: PbF_2 crystal was grown by the Bridgman technique for the first time. The spectroscopic properties, energy transfer mechanism, and first-principles calculations of as-grown crystals were investigated in detail. The co-doped Tm^{3+} ion can offer an appropriate sensitization and deactivation effect for Er^{3+} ion at the same time in PbF_2 crystal under the pump of conventional 800 nm laser diodes (LDs). With the introduction of Tm^{3+} ion into the Er^{3+}: PbF_2 crystal, the Er/Tm: PbF_2 crystal exhibited an enhancing 2.78 µm mid-infrared (MIR) emission. Furthermore, the cyclic energy transfer mechanism that contains several energy transfer processes and cross-relaxation processes was proposed, which would well achieve the population inversion between the Er^{3+}: $^4I_{11/2}$ and Er^{3+}: $^4I_{13/2}$ levels. First-principles calculations were performed to find that good performance originates from the uniform distribution of Er^{3+} and Tm^{3+} ions in PbF_2 crystal. This work will provide an avenue to design MIR laser materials with good performance.

Keywords: 2.78 µm mid-infrared emission; Er/Tm; PbF_2 laser crystal; energy transfer mechanism; first-principles calculation

1. Introduction

Over the past several decades, mid-infrared (MIR) solid-state lasers operating around 2.7–3 µm have received extensive attention for numerous applications in medicine surgery, communications, remote sensing, pollution monitoring, and military countermeasures, etc. [1–5]. Additionally, 2.7–3 µm lasers are suitable pump sources for longer wavelength mid-infrared or long-infrared (8–12 µm) laser applications utilizing the optical parametric oscillators [6,7].

Up to now, many kinds of rare-earth ions in favorable ~3 µm MIR emissions have been analyzed, such as erbium ion (Er^{3+}): $^4I_{11/2} \to {}^4I_{13/2}$ [8], holmium ion (Ho^{3+}): $^5I_6 \to {}^5I_7$ [9], and dysprosium ion (Dy^{3+}): $^6H_{13/2} \to {}^6H_{15/2}$ [10]. Among them, the Er^{3+} ion-doped single crystal has been deemed as the effective source for ~3 µm laser operation, benefiting from its abundant energy levels, such as GSGG [11], YSGG [12], YAP [13], Lu_2O_3 [14], $GdScO_3$ [15], SrF_2 crystals [16], $NdVO_4$ [17], $InVO_4$ crystals [18], etc. As investigated, the Er^{3+} ion can be directly pumped utilizing 808 or 980 nm commercial laser diodes (LDs) corresponding to Er^{3+} ion absorption transitions from ground state $^4I_{15/2}$ to $^4I_{9/2}$, $^4I_{11/2}$ levels, respectively. To take further advantage of this, a co-doping suitable sensitization ion having strong absorption around 980 nm or 800 nm would improve the absorption efficiency, such as Yb^{3+}, Nd^{3+}, or Tm^{3+} ions [19–21]. However, the fluorescence lifetime of the $^4I_{11/2}$ level (the upper) is fairly shorter than that of the $^4I_{13/2}$ level (the lower) of Er^{3+} ion, causing the possible termination of 2.7 μm mid-infrared emissions [22]. Therefore, the shortcoming of the intrinsic self-terminating "bottleneck" effect of Er^{3+} ion is important to consider. On one hand, the self-terminating "bottleneck" effect can be restrained by the energy transfer up-conversion (UC) process: $2\ ^4I_{13/2} \rightarrow\ ^4I_{15/2} +\ ^4I_{9/2}$, which needs heavy doping of Er^{3+} ion (>30 at.%). The UC process can simultaneously depopulate the $^4I_{13/2}$ level and populate the $^4I_{11/2}$ level via non-radiative transition from $^4I_{9/2}$ to $^4I_{11/2}$ levels. However, excessive Er^{3+} doping concentrations will generate the inclusion defects in as-grown crystal and degenerate the crystal quality and thermal performance, which is not conducive to laser output efficiency [23,24]. On the other hand, we can focus attention on co-doping a suitable deactivation ion for Er^{3+} ion to suppress the self-terminating effect, such as Pr^{3+}, Ho^{3+}, Dy^{3+}, or Tm^{3+} ions [25–28]. These deactivation ions can dramatically reduce the population of lower Er^{3+}: $^4I_{13/2}$ levels, thereby achieving efficient 2.7 μm mid-infrared emission. Based on the above investigation, it is noteworthy that Tm^{3+} ion can simultaneously serve as sensitization and deactivation effects for Er^{3+} ion [29,30].

In recent years, fluoride crystals have attracted numerous attention in the field of mid-infrared lasers, such as the β-PbF_2 crystal [31]. The PbF_2 crystal exhibits its intrinsic advantages. The PbF_2 crystal has lower phonon energy (257 cm^{-1}), compared with $GdLiF_4$ (432 cm^{-1}), $LiYF_4$ (442 cm^{-1}), $LuLiF_4$ (400 cm^{-1}) and BaY_2F_8 (415 cm^{-1}) crystals [32–34]. Such low phonon energy is conducive to reducing the non-radiative transition probability and enhancing the spontaneous radiation transition probability between $^4I_{11/2}$ and $^4I_{13/2}$ levels of Er^{3+} ion [35]. Moreover, the PbF_2 crystal is optically transparent in the region of 0.25–15 μm, which is broader than other fluoride crystals, such as $LiYF_4$ (0.12–8.0 μm), BaY_2F_8 (0.2–9.5 μm), and KYF_4 (0.15–9.0 μm). Additionally, another issue to consider is the physical properties of the material. Some fluoride crystals have low thermal conductivity, such as CaF_2 and SrF_2. The PbF_2 crystal has high thermal conductivity (28 W/m/K) and stable mechanical and chemical properties [36,37]. Consequently, with these favorable characteristics, the PbF_2 crystal may be selected as a promising host material.

In this paper, Er: PbF_2, Tm: PbF_2, Er/Tm: PbF_2 crystals were successfully prepared by the Bridgman technique. The spectroscopic properties of prepared crystals were analyzed based on absorption spectra, emission spectra, and fluorescence decay curves. Compared with the Er: PbF_2 crystal, the Er/Tm co-doped PbF_2 crystal presents a larger 2.78 μm fluorescence emission intensity and higher fluorescence branching ratio. Moreover, theoretical calculations were performed to discover that the co-doping of the Tm^{3+} ion can make the Er^{3+} and Tm^{3+} ions more evenly distributed in PbF_2 crystals, which can effectively break the local clusters of the Er^{3+} in Er: PbF_2 crystal, thus ensuring efficient energy transfer between Er^{3+} and Tm^{3+} ions, and resulting in the enhancement of 2.78 μm MIR fluorescence emission.

2. Experimental Section

The 1.0 at.% Er: PbF_2, 0.5 at.% Tm: PbF_2, and 1.0 at.% Er/0.5 at.% Tm: PbF_2 crystals were grown by the conventional Bridgman method in an atmosphere of N_2 with intermediate molybdenum heating. The fluoride powders of the PbF_2 (99.999%), ErF_3 (99.999%), and TmF_3 (99.999%) were all raw materials. The raw materials were weighed and thoroughly mixed. The process of crystal growth was similar to our previous work [37]. The melt was homogenized in a covered graphite crucible in a high-temperature zone at 1000 °C for 8 h, and the crystal growth process was driven by lowering the graphite crucible at a speed of 0.5 mm/h. After the growth process was completed, the cooling rate of the crystal was 30 °C/cm–40 °C/h. The actual concentration of Er^{3+} and Tm^{3+} ions in the grown samples were measured utilizing inductively coupled plasma atomic emission spectrometry (ICP-AES). The concentrations of Er^{3+} and Tm^{3+} ions in dual-doped Er/Tm: PbF_2 crystal were 1.15 at.%, and 0.58 at.%, respectively. The concentration of Er^{3+} ion in the Er: PbF_2 crystal was 1.15 at.%, and the concentration of Tm^{3+} ion in the Tm: PbF_2 crystal was 0.59 at.%.

The crystalline structure of as-grown samples was observed utilizing D/max2550 X-ray diffraction (XRD) with Cu K_α radiation. The Perkin–Elmer UV-VIS-NIR spectrometer (Lambda 900) with a resolution of 1 nm was used to detect the absorption spectra of prepared samples in the range of 400–2200 nm. The emission spectra, up-conversion fluorescence spectra, and fluorescence decay curves were detected and recorded using the Edinburgh Instruments FLS920 and FSP920 spectrophotometers. The repetition frequency of the excitation pulse for measuring the fluorescence decay curves was set to 20 Hz, and the duration of the excitation pulses was 30s. All the measurements were performed at room temperature.

3. Calculation Method

In the framework of density functional theory, VASP codes and the plane-wave basis set method were used for calculation [38,39]. The mutual interactions were described by the projector augmented-wave pseudopotential with an exchange-correlation function (Perdew–Burke–Ernzerhof form) [40,41]. The cut-off was set at 550 eV and a $1 \times 1 \times 1$ Gamma k-grid was used to guarantee the relaxation accuracy of 10^{-5} eV and 0.01 eVÅ^{-1} within a $2 \times 2 \times 2$ supercell, respectively. The spin polarization was included in the calculations. According to the method reported previously [42], the formation energy (ΔE) and cluster symbols were obtained. It is pointed out that the energy correction of the PbF_2 crystal was different from that of CaF_2, SrF_2, and BaF_2 crystals. For a $2 \times 2 \times 2$ supercell with a net charge, the calculated value in PbF_2 crystal was 0.069 eV.

4. Results and Discussion

4.1. Crystal Structure Analysis

Figure 1 shows the XRD patterns and refined XRD patterns of the Er: PbF_2, Tm: PbF_2, Er/Tm: PbF_2 crystals, and the JCPDS standard card of the PbF_2 crystal (nos. 06-2051) [37]. The residuals of refinements (fit profiles shown in Figure 1) of Er: PbF_2, Tm: PbF_2, Er/Tm: PbF_2 crystals were 9.61%, 7.71%, 10.17%, respectively. It is obvious that no clear shift in the phase diffraction peaks was observed and all XRD curves were well matched with the standard card of the β-PbF_2 crystalline phase (nos. 06-2051). The results demonstrate successful co-doping of Er^{3+} and Tm^{3+} ions in PbF_2 crystal without phase transitions.

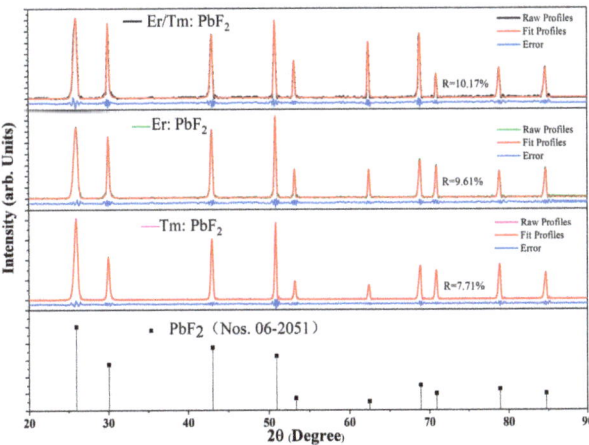

Figure 1. XRD patterns and refined XRD patterns of the Er/Tm: PbF_2, Er: PbF_2, Tm: PbF_2 crystals and PbF_2 crystal standard card.

4.2. First-Principles Calculations

Based on the first-principal calculations, the cluster structure of Tm^{3+} and Er^{3+} ions were simulated to research the change of local structures of doping ions in PbF_2 crystal. The possible thermodynamically stable Er^{3+} and Tm^{3+} centers in PbF_2 crystals are shown in Figure 2a,b. It is clear to see that there are 9 different types of centers in each Tm: PbF_2 and Er: PbF_2 crystals. In particular, only the $3_1|0|8|4_1$-C center in the Tm: PbF_2 crystal varies from the $2_1|0|6|3_1$ center in the Er: PbF_2 crystal, and the other eight different types of centers in the Tm: PbF_2 crystal are the same as the Er: PbF_2 crystal. Moreover, Figure 2c shows the formation energy of Er^{3+} and Tm^{3+} versus the number of Er^{3+} and Tm^{3+} ions within a cluster, respectively. It can be seen that the slope of Er^{3+} clusters in PbF_2 crystal is -0.988 eV, which is almost the same as the slope of Tm^{3+} clusters in the PbF_2 crystal (-1.003 eV). These results indicate that the clustering characteristics of Er^{3+} and Tm^{3+} ions in PbF_2 crystal are almost consistent. This phenomenon agreed well with the approximately equal segregator coefficients of Er^{3+} (1.15) and Tm^{3+} (1.16) in the Er/Tm: PbF_2 crystal mentioned above, which may be owing to the slightly different ion radii between Er^{3+} (88.1 pm) and Tm^{3+} (86.9 pm) ions. That is to say, it can be considered that the Er^{3+} and Tm^{3+} ions replace Pb^{2+} ions with equal probability when they are co-doped in PbF_2 crystal, which makes the Er^{3+} and Tm^{3+} ions more evenly distributed in the PbF_2 crystal. The results suggest that the efficient energy transfer between Er^{3+} and Tm^{3+} ions can be guaranteed due to the uniform distribution of Er^{3+} and Tm^{3+} ions, and result in the enhancing of 2.78 μm MIR fluorescence emission in the ensuing discussion.

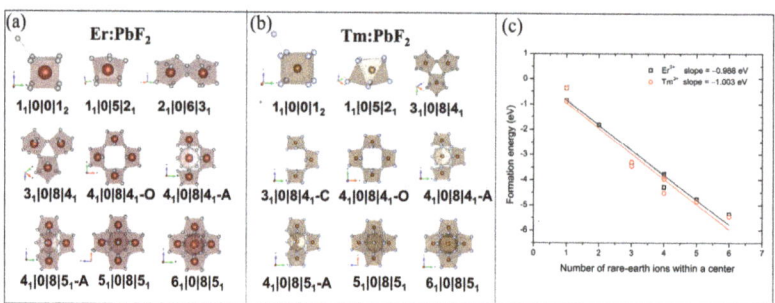

Figure 2. (a) Thermodynamically stable Er^{3+} centers in PbF_2 crystal; (b) Thermodynamically stable Tm^{3+} centers in PbF_2 crystal; (c) Formation energy of Er^{3+} and Tm^{3+} versus the number of rare-earth ions within a cluster.

4.3. Absorption Spectroscopy

The illustrations in Figure 3 shows the photos of Er/Tm: PbF_2, Er: PbF_2, Tm: PbF_2 crystals and their cut and polished crystal pieces; their sizes are also marked, respectively. It can be seen that all the crystal pieces are transparent and have no inclusions. Figure 3 illustrates the room temperature absorption spectra of Er: PbF_2, Tm: PbF_2, and Er/Tm: PbF_2 crystals ranging from 400 nm to 2200 nm. Clearly, the typical absorption bands centered at approximately 417, 451, 486, 521, 541, 650, 802, 975, and 1509 nm in the Er: PbF_2 crystal originated from the transitions from the ground state $^4I_{15/2}$ level to upper-lying $^2H_{9/2}$, $^4F_{5/2,3/2}$, $^4F_{7/2}$, $^2H_{11/2}$, $^4S_{3/2}$, $^4F_{9/2}$, $^4I_{9/2}$, $^4I_{11/2}$ and $^4I_{13/2}$ levels of Er^{3+} ion, respectively [37]. While in the Tm: PbF_2 crystal mainly five absorption bands of Tm^{3+} ion are labeled, the absorption peaks centered at round 464, 680, 792, 1211, and 1618 nm are in accord with the transitions from ground state $^3H_6$ level to upper-lying $^1G_4$, $^3F_{2,3}$, $^3H_4$, $^3H_5$ and $^3F_4$ levels, respectively. Obviously, the huge absorption band centered at around 792 nm in the range of 750–830 nm corresponding to Tm^{3+}: $^3H_6 \rightarrow {}^3H_4$ transition well coincides with the wavelength of 808 nm AlGaAs LD pumping. The absorption bands in the Er/Tm: PbF_2 crystal are altogether composed of the transitions of Er^{3+} and Tm^{3+} ions discussed above, indicating the successful introduction of both Er^{3+} and Tm^{3+}

ions. Strong overlap between the Tm^{3+}: $^3H_6 \rightarrow {}^3H_4$ absorption transition and the Er^{3+}: $^4I_{15/2} \rightarrow {}^4I_{9/2}$ absorption transition can be seen in the Er/Tm: PbF_2 crystal. The absorption overlap indicates that a possible nonradiative energy transfer process Tm^{3+}: $^3H_4 \rightarrow Er^{3+}$: $^4I_{9/2}$ would effectively occur for enhancing the absorption efficiency of Er^{3+} ion ~800 nm. Therefore, benefiting from the broad absorption band of Tm^{3+} ion centered at around LD pump wavelength and the possibility for energy transfer, the Tm^{3+} ion can act as a suitable sensitizer for Er^{3+} ion in the Er/Tm dual-doped PbF_2 crystal.

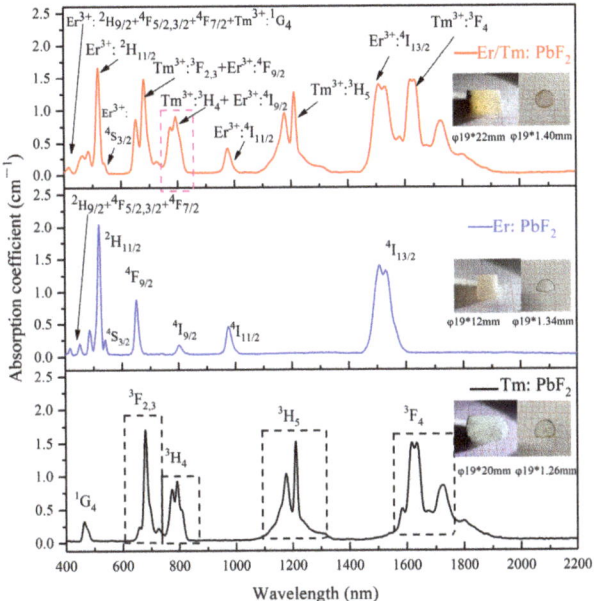

Figure 3. Absorption spectra of Tm: PbF_2, Er: PbF_2, and Er/Tm: PbF_2 crystals ranging from 400 to 2200 nm at room temperature. (Illustration: the photos of Er/Tm: PbF_2, Er: PbF_2, Tm: PbF_2 crystals and their cut and polished crystal pieces, respectively.)

For demonstrating the sensitization effect of Tm^{3+} ion for Er^{3+} ion via the Tm^{3+}: $^3H_4 \rightarrow Er^{3+}$: $^4I_{9/2}$ energy transfer transition, the lifetimes of Tm^{3+}: $^3H_4$ level in the Tm^{3+} single-doped and Er/Tm dual-doped PbF_2 crystals were measured and shown in Figure 4a,b, respectively. The decay curves were measured under the condition of 1.47 μm emission (Tm^{3+}: $^3H_4 \rightarrow {}^3F_4$) and 800 nm excitation (Tm^{3+}: $^3H_6 \rightarrow {}^3H_4$) and were all well fitted by single-exponential behavior. As shown in Figure 4a, the measured lifetime of the Tm^{3+}: $^3H_4$ manifold is 1.67 ms in the Tm: PbF_2 crystal, while the lifetime is 0.54 ms in the Er/Tm: PbF_2 crystal shown in Figure 4b. The remarkable decreasing lifetime in the Er/Tm: PbF_2 crystal indicates the effective sensitization effect of the Tm^{3+} ion. The energy transfer efficiency from Tm^{3+}: $^3H_4$ to Er^{3+}: $^4I_{9/2}$ level can be calculated by the following equation: $\eta_{ET1} = 1 - \tau_{Er/Tm}/\tau_{Tm}$, where $\tau_{Er/Tm}$ and τ_{Tm} are the lifetimes of Tm^{3+}: $^3H_4$ level in Tm: PbF_2, Er/Tm: PbF_2 crystals, respectively. The high value of η_{ET1} (67.66%) confirms that the Tm^{3+} ion has a significant influence on Er^{3+}: $^4I_{9/2}$ level in PbF_2 crystal, and can effectively act as a sensitizer for Er^{3+} ion for enhancing ~2.7 μm MIR emission.

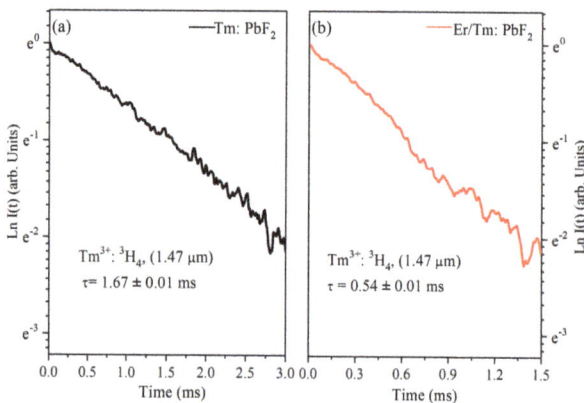

Figure 4. (a) Fluorescence decay curves of the Tm^{3+}: $^3H_4$ energy level of Tm: PbF_2 crystal; (b) fluorescence decay curves of the Tm^{3+}: $^3H_4$ energy level of Er/Tm: PbF_2 crystal (λ_{ex} = 800 nm, λ_{em} = 1470 nm).

4.4. Emission Spectra and Emission Cross-Sections

For further clarifying the energy transfer mechanism between Tm^{3+} and Er^{3+} ions, the emission spectra of Er/Tm: PbF_2, Er: PbF_2 samples in the range of 1400–1700 nm, and Er/Tm: PbF_2, Tm: PbF_2 samples in the 1700–2200 nm region are shown in Figure 5a,b, respectively. The test parameters of the luminescence performance of the prepared samples, such as pump power and slits, are uniformed. As shown in Figure 5a, compared with the Er: PbF_2 crystal the emission intensity centered at around 1.55 μm corresponding to the Er^{3+}: $^4I_{13/2} \rightarrow {}^4I_{15/2}$ transition in the Er/Tm: PbF_2 crystal weakened sharply, at almost ten times lower. The result shows that the introduction of Tm^{3+} ion would significantly reduce the population of the Er^{3+}: $^4I_{13/2}$ energy level, thereby enhancing the ~2.7 μm mid-infrared emission and reversely weakening the 1.55 μm infrared emission. This depopulation of Er^{3+}: $^4I_{13/2}$ energy level is mainly attributed to the deactivation effect of Tm^{3+} ions via energy transfer process: Er^{3+}: $^4I_{13/2} \rightarrow Tm^{3+}$: $^3F_4$ in Er/Tm: PbF_2 crystal. As the deactivation energy transfer process occurs, the population on the Tm^{3+}: $^3F_4$ level would increase, thereby enhancing the 1.91 μm emission (Tm^{3+}: $^3F_4 \rightarrow {}^3H_6$ transition) in the Er/Tm: PbF_2 crystal, but it is actually weakened (shown in Figure 5b). The 1.91 μm emission intensity of the Tm^{3+} ion in Er/Tm: PbF_2 crystal is nearly three times lower than that in the Tm^{3+} single doped PbF_2 crystal. This result is mainly assigned to the cross-relaxation (CR) process between Tm^{3+} and Er^{3+} ions (Tm^{3+}: $^3F_4 + Er^{3+}$: $^4I_{13/2} \rightarrow Tm^{3+}$: $^3H_4 + Er^{3+}$: $^4I_{15/2}$), bringing about the depopulation of the Tm^{3+}: $^3F_4$ level and Er^{3+}: $^4I_{13/2}$ level. Therefore, the reduced emission intensity of 1.55 μm of Er^{3+} ion and 1.91 μm of Tm^{3+} ion both would depopulate the ions on the Er^{3+}: $^4I_{13/2}$ level, which is beneficial to enhance ~2.7 μm MIR emission. More importantly, as shown in Figure 6, the emission intensity of the Er/Tm: PbF_2 crystal centered at around 2.7 μm in the 2500–3100 nm region is remarkably larger than that of the Er: PbF_2 crystal, confirming that the efficient enhanced ~2.7 μm emission is achieved in the Er/Tm: PbF_2 designed crystal. To further confirm the prospects of Er: PbF_2, Er/Tm: PbF_2 crystals as the mid-infrared luminescent material in laser applications, the 2.78 μm emission cross-sections are subsequently calculated according to the Fuchtbauere–Ladenburg theory [43]:

$$\sigma_{em}(\lambda) = \frac{A\beta\lambda^5 I(\lambda)}{8\pi c n^2 \int \lambda I(\lambda) d\lambda} \quad (1)$$

where λ denotes the wavelength of fluorescence spectrum, $I(\lambda)$ is the intensity of emission spectrum at λ, $I(\lambda)/\int \lambda I(\lambda)d\lambda$ is the normalized line shape function of the emission spectrum of prepared crystal, n is the refractive index of PbF_2 crystal, c is the speed of light in a vacuum, β is the fluorescence branching ratio of $^4I_{11/2} \rightarrow {}^4I_{13/2}$ transition, and A is the spontaneous emission probability. The value of β for ~2.7 μm mid-infrared emission in Er: PbF$_2$ is calculated to be 14.9%, and in the Er/Tm: PbF$_2$ crystal is calculated to be 20.24%. The maximum emission cross-section of the Er/Tm: PbF$_2$ crystal is calculated to be 0.63×10^{-20} cm^2 at 2780 nm, which is almost twice that of the Er: PbF$_2$ crystal (0.32×10^{-20} cm^2). Moreover, as shown in Table 1, this higher stimulated emission cross-section in the Er/Tm: PbF$_2$ crystal possibly coincides well with the higher fluorescence branching ratio β (20.24%) of the Er^{3+}: $^4I_{11/2} \rightarrow {}^4I_{13/2}$ transition. A higher emission cross-section is more favorable in achieving high performance of MIR laser operation. These results are related to the more uniform distribution of Er^{3+} and Tm^{3+} ions in PbF$_2$ crystal after the co-doping of Tm^{3+} ions, which is consistent with the theoretical calculation results. Furthermore, it is pointed out that the enhancing of 2.78 μm MIR fluorescence emission is more dependent on the efficient energy transfer between Er^{3+} and Tm^{3+} ions, which comes from the uniform distribution of doped ions.

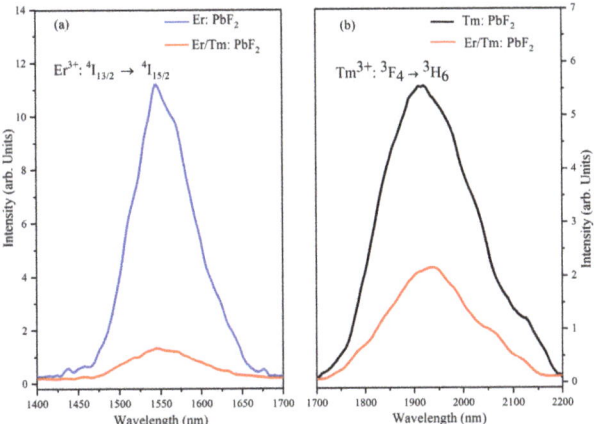

Figure 5. (**a**) Emission spectra of the Er: PbF$_2$, Er/Tm: PbF$_2$ crystals in the range of 1400–1700 nm (λ_{ex} = 800nm); (**b**) emission spectra of Tm: PbF$_2$, Er/Tm: PbF$_2$ crystals in the range of 1700–2200 nm (λ_{ex} = 800nm).

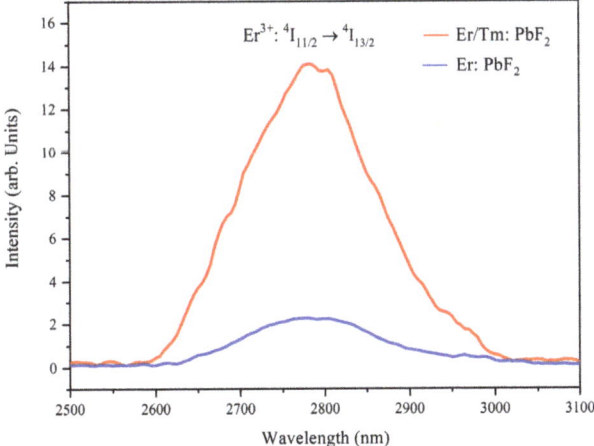

Figure 6. Emission spectra of the Er: PbF$_2$, Er/Tm: PbF$_2$ crystals in the range of 2500–3100 nm (λ_{ex} = 800 nm).

Table 1. MIR emission cross-sections σ_{em}, and lifetimes of $^4I_{11/2}$, $^4I_{13/2}$ levels of Er/Tm: PbF$_2$, Er: PbF$_2$ crystals compared with other Er^{3+}-doped crystals.

Crystal	σ_{em} (10^{-20} cm^2)	$\tau(^4I_{11/2})$ (ms)	$\tau(^4I_{13/2})$ (ms)	$\tau(^4I_{11/2})/\tau(^4I_{13/2})$ (%)	Ref.
1.0 at.% /0.5 at.% Er/Tm: PbF$_2$	0.63@2780nm	6.91 ± 0.01	3.14 ± 0.01	220.06	[This work]
1 at.% Er: PbF$_2$	0.32@2780nm	6.03 ± 0.01	12.06 ± 0.05	50.00	[This work]
10 at.% Er:BaLaGa$_3$O$_7$	7.34@2714nm	0.72	7.99	9.01	[44]
10 at.% Er:CaLaGa$_3$O$_7$	17.9@2702nm	0.77	8.41	9.16	[45]
8 at.%Er:LuAl$_3$(BO$_3$)$_4$	8.60@3170nm	2.10	2.54	82.68	[46]
7 at.% Er:Y$_2$O$_3$	1.41@2723nm	2.95	17.57	16.79	[47]
10 at.% Er:SrGdGa$_3$O$_7$	1.30@2.7μm	1.10	4.48	24.55	[24]
5at.% Er:YAP	9.00@2792nm	0.85	7.30	11.64	[13]
7 at.% Er:Lu$_2$O$_3$	1.10@2730nm	1.10	4.30	25.58	[14]
5 at.% Er:GdScO$_3$	0.93@2720nm	2.24	4.57	49.02	[15]
4 at.% Er:SrF$_2$	0.78@2727nm	9.56	15.06	63.48	[16]
4 at.% Er:CaF$_2$	0.65@2720nm	5.98	9.94	60.16	

4.5. Energy Transfer Mechanism between Tm^{3+} and Er^{3+} Ions

Based on spectroscopic results discussed above, the simplified energy level scheme and electron transitions of the Er^{3+}/Tm^{3+} co-doped PbF$_2$ crystal are presented in Figure 7. The cyclic related processes of the Tm^{3+} and Er^{3+} ions in the crystal under optical excitation are as follows: cross-relaxation, energy transfer between Tm^{3+} and Er^{3+} ions, and multiphonon relaxation. The main two ET (namely ET1, ET2) and three CR (namely CR1, CR2, CR3) processes are listed as follows:

ET 1: Tm^{3+}: $^3H_4$ + Er^{3+}: $^4I_{15/2}$ → Tm^{3+}: $^3H_6$ + Er^{3+}: $^4I_{9/2}$;
ET 2: Er^{3+}: $^4I_{13/2}$ + Tm^{3+}: $^3H_6$ → Er^{3+}: $^4I_{15/2}$ + Tm^{3+}: $^3F_4$;
CR 1: Tm^{3+}: $^3F_4$ + Er^{3+}: $^4I_{13/2}$ → Tm^{3+}: $^3H_6$ + Er^{3+}: $^4I_{9/2}$;
CR 2: Tm^{3+}: $^3F_4$ + Er^{3+}: $^4I_{13/2}$ → Tm^{3+}: $^3H_4$ + Er^{3+}: $^4I_{15/2}$;
CR 3: 2Er^{3+}: $^4I_{13/2}$ → Er^{3+}: $^4I_{15/2}$ + Er^{3+}: $^4I_{9/2}$.

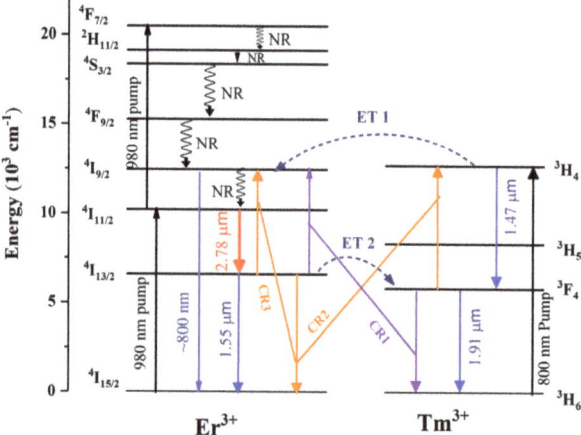

Figure 7. Simplified energy level scheme and electron transitions of Er^{3+}/Tm^{3+} co-doped system.

As discussed, the Tm^{3+}: $^3H_4$ → $^3H_6$ transition is resonant with the Er^{3+}: $^4I_{15/2}$ → $^4I_{9/2}$ transition in the Er/Tm: PbF$_2$ crystal. Therefore, after the crystal is excited to the Tm^{3+}: $^3H_4$ level by a pump of 800 nm LD, ET1 process Tm^{3+}: $^3H_4$ → Er^{3+}: $^4I_{9/2}$ would occur. Ions in the Er^{3+}: $^4I_{9/2}$ level decay non-radiatively to the lower Er^{3+}: $^4I_{11/2}$ level, and then decay radiatively to the Er^{3+}: $^4I_{13/2}$ level and emit 2.78 μm mid-infrared light. Ions in the Er^{3+}:

$^4I_{13/2}$ level continue to decay radiatively to the ground state Er^{3+}: $^4I_{15/2}$ level and emit 1.55 μm infrared light. Similarly, the Er^{3+}: $^4I_{13/2} \rightarrow {}^4I_{15/2}$ transition is resonant with the Tm^{3+}: $^3H_6 \rightarrow {}^3F_4$ transition, and the ET2 process from Er^{3+}: $^4I_{13/2}$ to Tm^{3+}: $^3F_4$ level takes place. The ET2 process would reduce the population of the lower level of Er^{3+}: $^4I_{13/2}$, thereby enhancing the 2.78 μm emission and weakening the 1.55 μm emission, as shown in Figures 5a and 6. Meantime, the energy transfer up-conversion (UC) CR3 process (Er^{3+}: $2{}^4I_{13/2} \rightarrow {}^4I_{15/2} + {}^4I_{9/2}$) in the crystal can also populate the Er^{3+}: $^4I_{11/2}$ level and depopulate the Er^{3+}: $^4I_{13/2}$ level. Additionally, ions in the Tm^{3+}: $^3F_4$ level decay radiatively to the $^3H_6$ level and emit 1.91 μm emission. The subsequent CR1 populates the Er^{3+}: $^4I_{9/2}$ level, and then the Er^{3+}: $^4I_{11/2}$ level is populated through the nonradiative decay from the $^4I_{9/2}$ level to the $^4I_{11/2}$ level, increasing the population ratio of $^4I_{11/2}/{}^4I_{13/2}$ levels. Moreover, the ions in the Tm^{3+}: $^3F_4$ energy level will also absorb energy and jump to the upper Tm^{3+}: $^3H_4$ energy level due to Stark level splitting, and then the CR2 process described above occurs. The CR2 process can simultaneously reduce the population Er^{3+}: $^4I_{13/2}$, Tm^{3+}: $^3F_4$ levels, to achieve 2.78 μm emission enhancement and 1.91 μm emission reduction, as shown in Figures 5b and 6. The CR2 process also brings about the increasing population of the Tm^{3+}: $^3H_4$ level. Besides emitting 1.47 μm light via the Tm^{3+}: $^3H_4 \rightarrow {}^3F_4$ transition, ions in the Tm^{3+}: $^3H_4$ level can populate the Er^{3+}: $^4I_{9/2}$ level via ET1 process, resulting in further enhancement of the sensitization effect. To prove the CR2 process, the UC emission spectra of Er: PbF_2 and Er/Tm: PbF_2 crystals are shown in Figure 8 under 980 nm excitation. Clearly, as shown in Figure 7, under 980 nm NIR light excitation, the electrons in the ground level $^4I_{15/2}$ can be excited to the intermediate level $^4I_{11/2}$, and the electrons in the $^4I_{11/2}$ level sequentially populate the $^4F_{7/2}$ level ($^4I_{15/2} \rightarrow {}^4I_{11/2} \rightarrow {}^4F_{7/2}$). Additionally, then, the multiple nonradiative multi-phonon relaxation in the $^4F_{7/2}$ state in turn populate the lower $^2H_{11/2}$, $^4S_{3/2}$, $^4F_{9/2}$, and $^4I_{9/2}$ levels, which would produce 800 nm light via the process: $^4I_{9/2} \rightarrow {}^4I_{15/2}$. It is clear to see that the UC emission intensity of the Er/Tm: PbF_2 crystal is at least two times larger than that of the Er: PbF_2 crystal at around 800 nm. Obviously, Tm^{3+} ions have no absorption band matching the 980 nm excitation (shown in Figure 3). This enhancing UC emissions phenomenon is possibly assigned to the CR2 and ET1 mechanism processes illustrated in Figure 7. To summarize, the ET1, ET2, CR1, CR2, CR3 processes all have significant effects on narrowing the lifetime gap of upper-lying Er^{3+}: $^4I_{11/2}$ and lower-lying Er^{3+}: $^4I_{13/2}$ levels or even achieving population conversion of these two levels, thereby obtaining efficient enhanced 2.78 μm emission.

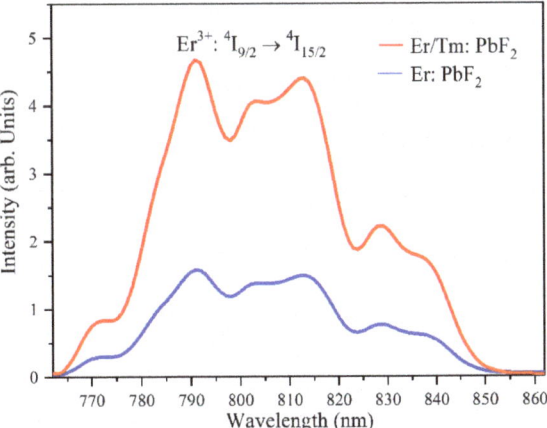

Figure 8. Up-conversion emissions of Er: PbF_2, and Er/Tm: PbF_2 crystals in the range of 760–860 nm (λ_{ex} = 980nm).

4.6. Fluorescence Decay Curves and Fluorescence Lifetimes

For further demonstrating the energy interaction mechanism between Er^{3+} and Tm^{3+} ions, the time-resolved decay curves of the Er^{3+} ion 2.78 μm ($^4I_{11/2} \to {}^4I_{13/2}$) and 1.55 μm ($^4I_{13/2} \to {}^4I_{13/2}$) fluorescence emission for the Er/Tm: PbF$_2$ and Er: PbF$_2$ crystals were measured and shown in Figure 9. The lifetimes of $^4I_{13/2}$ levels were measured under the conditions of 1.55 μm emission ($^4I_{13/2} \to {}^4I_{15/2}$) and 1.49 μm excitation ($^4I_{15/2} \to {}^4I_{13/2}$). The decay curves of the Er^{3+}: $^4I_{11/2}$ and Er^{3+}: $^4I_{13/2}$ levels are well fitted with single-exponential behavior. As shown in Figure 9a,b, the measured lifetime of the upper-lying $^4I_{11/2}$ level in the Er/Tm: PbF$_2$ crystal (6.91 ms) is 14.6% longer compared with the Er: PbF$_2$ crystal (6.03 ms), which is assigned to the sensitization effect of the Tm^{3+} ion on the upper-lying Er^{3+}: $^4I_{11/2}$ level. Moreover, as shown in Figure 9c,d, the measured lifetime of the lower-lying $^4I_{13/2}$ level in the Er/Tm: PbF$_2$ crystal is 3.14 ms, which is 73.96% shorter compared with the Er: PbF$_2$ crystal (12.06 ms). This remarkable decrease of the lifetime of lower-lying $^4I_{13/2}$ level denotes that Tm^{3+} ions can dramatically depopulate the Er^{3+}: $^4I_{13/2}$ level via ET2, CR1, CR2, CR3 processes, thereby enhancing the 2.78 μm emission in PbF$_2$ crystals. The ET2, CR1, CR2, CR3 processes all have significant effects on narrowing the lifetime gap of upper-lying Er^{3+}: $^4I_{11/2}$ and lower-lying Er^{3+}: $^4I_{13/2}$ levels or even achieving population conversion of these two levels. Besides, the energy transfer efficiency η_{ET2} was calculated to be 73.96%, confirming the efficient deactivation effect of the Tm^{3+} ion for the Er^{3+} ion. Furthermore, Table 1 shows the lifetimes of $^4I_{11/2}$, $^4I_{13/2}$ levels of Er/Tm: PbF$_2$, Er: PbF$_2$ crystals, and other Er^{3+} doped laser crystals. The shorter fluorescence lifetime of $^4I_{13/2}$ lower level induces the longer fluorescence lifetime ratio $\tau(^4I_{11/2})/\tau(^4I_{13/2})$. The fluorescence lifetime ratio $\tau(^4I_{11/2})/\tau(^4I_{13/2})$ in Er/Tm: PbF$_2$ crystal is 220.06%, which is dramatically larger than that of the Er: PbF$_2$ crystal (50.00%) and other Er^{3+} doped crystals. The remarkably enhanced $\tau(^4I_{11/2})/\tau(^4I_{13/2})$ ratio in Er/Tm: PbF$_2$ crystal is favorable for achieving efficient laser operation ~2.7 μm. As a consequence, the introduction of Tm^{3+} ions can simultaneously act as sensitization and deactivation ions for the Er^{3+} ion, thereby enhancing 2.78 μm mid-infrared emission and reducing the laser threshold of 2.78 μm luminescence.

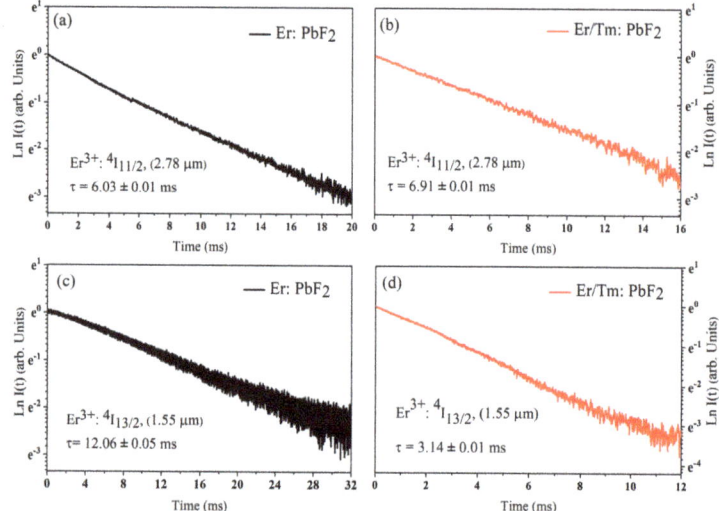

Figure 9. (a) Fluorescence decay curves of the Er^{3+}: $^4I_{11/2}$ energy level of Er: PbF$_2$ crystal (λ_{ex} = 800 nm, λ_{em} = 2780 nm); (b) Er^{3+}: $^4I_{11/2}$ energy level of Er/Tm: PbF$_2$ crystal (λ_{ex} = 800 nm, λ_{em} = 2780 nm); (c) Er^{3+}: $^4I_{13/2}$ energy level of Er: PbF$_2$ crystal (λ_{ex} = 1490 nm, λ_{em} = 1550 nm); (d) Er^{3+}: $^4I_{13/2}$ energy level of Er/Tm: PbF$_2$ crystal (λ_{ex} = 1490 nm, λ_{em} = 1550 nm).

5. Conclusions

In summary, Er^{3+}: PbF_2, Tm^{3+}: PbF_2, and Er^{3+}/Tm^{3+}: PbF_2 crystals were prepared successfully by the Bridgman technique. An efficient enhanced 2.78 μm emission was obtained in the Er/Tm: PbF_2 crystal for the first time, and the proposed energy transfer mechanism of the Er/Tm: PbF_2 crystal was systematically investigated. The theoretical calculations were performed to discover that the co-doping of Tm^{3+} ions can make the Er^{3+} and Tm^{3+} ions more evenly distributed in PbF_2 crystals, which can effectively break the local clusters of Er^{3+} in Er: PbF_2 crystal, thus ensuring efficient energy transfer between Er^{3+} and Tm^{3+} ions, and resulting in the enhancing of 2.78 μm MIR fluorescence emission. The cyclic energy transfer mechanism contains several energy transfer processes and cross-relaxation processes, which all have significant effects on narrowing the lifetime gap of upper-lying Er^{3+}: $^4I_{11/2}$ and lower-lying Er^{3+}: $^4I_{13/2}$ levels or even achieving population conversion of these two levels. As proved, the Tm^{3+} ion can simultaneously act as an appropriate sensitized and deactivated ion for the Er^{3+} ion in the PbF_2 crystal. Compared with the Er^{3+} single-doped crystal, the Er^{3+}/Tm^{3+} co-doped PbF_2 crystal has the larger 2.78 μm mid-infrared fluorescence emission intensity, higher fluorescence branching ratio (20.24%), and higher stimulated emission cross-section (0.63×10^{-20} cm^2), corresponding to Er^{3+}: $^4I_{11/2} \to {}^4I_{13/2}$ transition. Therefore, the introduction of Tm^{3+} ions is favorable for achieving efficient enhanced 2.78 μm emission in the Er/Tm: PbF_2 crystal, which can become a promising material for low threshold, and high-efficiency mid-infrared laser applications under the pump of a conventional 800 nm LD.

Author Contributions: Conceptualization, P.Z.; methodology, J.L. and Q.C.; software, F.M.; validation, Y.H.; formal analysis, S.Z.; investigation, J.L., H.T. and Q.Y.; resources, Y.H.; and Z.L.; data curation, J.L. and Q.C.; writing—original draft preparation, J.L. and X.N.; writing—review and editing, P.Z.; visualization, Q.C.; supervision, Z.C.; project administration, Z.C.; funding acquisition, Y.H. and Z.L. All authors have read and agreed to the published version of the manuscript.

Funding: This work was supported by the National Natural Science Foundation of China (NSFC) (51972149, 51872307, 61935010, 51702124); Key-Area Research and Development Program of Guangdong Province (2020B090922006); Guangdong Project of Science and Technology Grants (2018B0303230 17, 2018B010114002); Guangzhou science and technology project (201904010385, 201903010042); The Fundamental Research Funds for the Central Universities (21620445).

Institutional Review Board Statement: Not applicable.

Informed Consent Statement: Not applicable.

Data Availability Statement: The data presented in this study are available on request from the corresponding author.

Conflicts of Interest: The authors declare no conflict of interest.

References

1. Popov, A.; Sherstnev, V.; Yakovlev, Y.P.; Baranov, A.; Alibert, C. Powerful Mid-Infrared Light Emitting Diodes for Pollution Monitoring. *Electron. Lett.* **1997**, *33*, 86–88. [CrossRef]
2. Bekman, H.T.; Van Den Heuvel, J.; Van Putten, F.; Schleijpen, R. Development of a Mid-Infrared Laser for Study of Infrared Countermeasures Techniques. In Proceedings of the Technologies for Optical Countermeasures, London, UK, 26–27 October 2004; Volume 5615, pp. 27–38.
3. Hu, T.; Dong, B.; Luo, X.; Liow, T.Y.; Song, J.; Lee, C.; Lo, G.Q. Silicon Photonic Platforms for Mid-Infrared Applications. *Photonics Res.* **2017**, *5*, 417–430. [CrossRef]
4. Walsh, B.M.; Lee, H.R.; Barnes, N.P. Mid Infrared Lasers for Remote Sensing Applications. *J. Lumin.* **2016**, *169*, 400–405. [CrossRef]
5. Urich, A.; Maier, R.R.J.; Yu, F.; Knight, J.C.; Hand, D.P. Silica Hollow Core Microstructured Fibres for Mid-Infrared Surgical Applications. *J. Non-Cryst. Solids* **2013**, *377*, 236–239. [CrossRef]
6. Godard, A. Infrared (2–12 Mm) Solid-State Laser Sources: A Review. *Comptes Rendus Phys.* **2007**, *8*, 1100–1128. [CrossRef]
7. Vodopyanov, K.; Ganikhanov, F.; Maffetone, J.; Zwieback, I.; Ruderman, W. $ZnGeP_2$ Optical Parametric Oscillator with 3.8–12.4-μm Tunability. *Opt. Lett.* **2000**, *25*, 841–843. [CrossRef]
8. Zhang, T.; Feng, G.; Zhang, H.; Yang, X.; Dai, S.; Zhou, S. 2.78 μm Passively Q-Switched Er^{3+}-Doped Zblan Fiber Laser Based on PLD-Fe^{2+}: Znse Film. *Laser Phys. Lett.* **2016**, *13*, 075102. [CrossRef]

9. Zhang, C.; Hao, Q.; Zu, Y.; Zong, M.; Guo, J.; Zhang, F.; Ge, Y.; Liu, J. Graphdiyne Saturable Absorber for Passively Q-Switched Ho^{3+}-Doped Laser. *Nanomaterials* **2020**, *10*, 1848. [CrossRef]
10. Pajewski, L.; Sójka, L.; Lamrini, S.; Benson, T.M.; Seddon, A.B.; Sujecki, S. Gain-Switched Dy^{3+}: Zblan Fiber Laser Operating around 3 µm. *J. Phys. Photonics* **2019**, *2*, 014003. [CrossRef]
11. Wu, Z.; Sun, D.; Wang, S.; Luo, J.; Li, X.; Huang, L.; Hu, A.; Tang, Y.; Guo, Q. Performance of a 967 nm Cw Diode End-Pumped Er: GSGG Laser at 2.79 µm. *Laser Phys.* **2013**, *23*, 055801. [CrossRef]
12. Arbabzadah, E.; Chard, S.; Amrania, H.; Phillips, C.; Damzen, M. Comparison of a Diode Pumped Er: YSGG and Er: YAG Laser in the Bounce Geometry at the 3 µm Transition. *Opt. Express* **2011**, *19*, 25860–25865. [CrossRef]
13. Kawase, H.; Yasuhara, R. 2.92-µm High-Efficiency Continuous-Wave Laser Operation of Diode-Pumped Er: YAP Crystal at Room Temperature. *Opt. Express* **2019**, *27*, 12213–12220. [CrossRef]
14. Li, T.; Beil, K.; Kränkel, C.; Huber, G. Efficient High-Power Continuous Wave Er: Lu$_2$O$_3$ Laser at 2.85 µm. *Opt. Lett.* **2012**, *37*, 2568–2570. [CrossRef] [PubMed]
15. Hou, W.; Zhao, H.; Qin, Z.; Liu, J.; Wang, D.; Xue, Y.; Wang, Q.; Xie, G.; Xu, X.; Xu, J. Spectroscopic and Continuous-Wave Laser Properties of Er: GdScO$_3$ Crystal at 2.7 µm. *Opt. Mater. Express* **2020**, *10*, 2730–2737. [CrossRef]
16. Ma, W.; Qian, X.; Wang, J.; Liu, J.; Fan, X.; Liu, J.; Su, L.; Xu, J. Highly Efficient Dual-Wavelength Mid-Infrared Cw Laser in Diode End-Pumped Er: SrF$_2$ Single Crystals. *Sci. Rep.* **2016**, *6*, 1–7. [CrossRef]
17. Bandiello, E.; Sánchez-Martín, J.; Errandonea, D.; Bettinelli, M. Pressure effects on the optical properties of NdVO$_4$. *Crystals* **2019**, *9*, 237. [CrossRef]
18. Botella, P.; Enrichi, F.; Vomiero, A.; Munõz-Santiuste, J.E.; Garg, A.B.; Arvind, A.; Manjón, F.J.; Segura, A.; Errandonea, D. Investigation on the Luminescence Properties of InMO$_4$ (M = V^{5+}, Nb^{5+}, Ta^{5+}) Crystals Doped with Tb^{3+} or Yb^{3+} Rare Earth Ions. *ACS Omega* **2020**, *5*, 2148–2158. [CrossRef] [PubMed]
19. Zhang, P.; Chen, Z.; Hang, Y.; Li, Z.; Yin, H.; Zhu, S.; Fu, S.; Li, A. Enhanced 2.7 µm Mid-Infrared Emissions of Er^{3+} Via Pr^{3+} Deactivation and Yb^{3+} Sensitization in LiNbO$_3$ Crystal. *Opt. Express* **2016**, *24*, 25202–25210. [CrossRef]
20. Li, S.; Zhang, L.; Zhang, P.; Hang, Y. Nd^{3+} as Effective Sensitization and Deactivation Ions in Nd, Er: LaF$_3$ Crystal for the 2.7 µm Lasers. *J. Alloy. Compd.* **2020**, *827*, 154268. [CrossRef]
21. Li, M.; Liu, X.; Guo, Y.; Hao, W.; Hu, L.; Zhang, J. ~2 µm Fluorescence Radiative Dynamics and Energy Transfer between Er^{3+} and Tm^{3+} Ions in Silicate Glass. *Mater. Res. Bull.* **2014**, *51*, 263–270. [CrossRef]
22. Sandrock, T.; Diening, A.; Huber, G. Laser Emission of Erbium-Doped Fluoride Bulk Glasses in the Spectral Range from 2.7 to 2.8 µm. *Opt. Lett.* **1999**, *24*, 382–384. [CrossRef] [PubMed]
23. Wang, J.; Cheng, T.; Wang, L.; Yang, J.; Sun, D.; Yin, S.; Wu, X.; Jiang, H. Compensation of Strong Thermal Lensing in an LD Side-Pumped High-Power Er: YSGG Laser. *Laser Phys. Lett.* **2015**, *12*, 105004. [CrossRef]
24. Gao, S.; Chen, T.; Hu, M.; Xu, S.; Xiong, Y.; Cheng, S.; Zhang, W.; Wang, Y.; Yang, W. Effects of Er^{3+} Concentration on the Optical Properties of Er^{3+}: SrGdGa$_3$O$_7$ Single Crystals. *Opt. Mater.* **2019**, *98*, 109502. [CrossRef]
25. Quan, C.; Sun, D.; Luo, J.; Zhang, H.; Fang, Z.; Zhao, X.; Hu, L.; Cheng, M.; Zhang, Q.; Yin, S. Growth, Structure and Spectroscopic Properties of Er, Pr: YAP Laser Crystal. *Opt. Mater.* **2018**, *84*, 59–65. [CrossRef]
26. Kang, S.; Yu, H.; Ouyang, T.; Chen, Q.; Huang, X.; Chen, Z.; Qiu, J.; Dong, G. Novel Er^{3+}/Ho^{3+}-Codoped Glass-Ceramic Fibers for Broadband Tunable Mid-Infrared Fiber Lasers. *J. Am. Ceram. Soc.* **2018**, *101*, 3956–3967. [CrossRef]
27. Wang, T.; Huang, F.; Ren, G.; Cao, W.; Tian, Y.; Lei, R.; Zhang, J.; Xu, S. Broadband 2.9 µm Emission and High Energy Transfer Efficiency in Er^{3+}/Dy^{3+} Co-Doped Fluoroaluminate Glass. *Opt. Mater.* **2018**, *75*, 875–879. [CrossRef]
28. Tian, Y.; Li, B.; Wang, J.; Liu, Q.; Chen, Y.; Zhang, J.; Xu, S. The Mid-Infrared Emission Properties and Energy Transfer of Tm^{3+}/Er^{3+} Co-Doped Tellurite Glass Pumped by 808/980 nm Laser Diodes. *J. Lumin.* **2019**, *214*, 116586. [CrossRef]
29. Chai, G.; Dong, G.; Qiu, J.; Zhang, Q.; Yang, Z. 2.7 µm Emission from Transparent Er^{3+}, Tm^{3+} Codoped Yttrium Aluminum Garnet (Y$_3$Al$_5$O$_{12}$) Nanocrystals–Tellurate Glass Composites by Novel Comelting Technology. *J. Phys. Chem. C* **2012**, *116*, 19941–19950. [CrossRef]
30. Tian, Y.; Xu, R.; Hu, L.; Zhang, J. 2.7 µm Fluorescence Radiative Dynamics and Energy Transfer between Er^{3+} and Tm^{3+} Ions in Fluoride Glass under 800 Nm and 980 Nm Excitation. *J. Quant. Spectrosc. Radiat.* **2012**, *113*, 87–95. [CrossRef]
31. Cazorla, C.; Errandonea, D. Superionicity and polymorphism in calcium fluoride at high pressure. *Phys. Rev. Lett.* **2014**, *113*, 235902. [CrossRef]
32. Zhang, X.; Schulte, A.; Chai, B. Raman Spectroscopic Evidence for Isomorphous Structure of GdLiF$_4$ and YLiF$_4$ Laser Crystals. *Solid State Commun.* **1994**, *89*, 181–184. [CrossRef]
33. Zhao, C.; Hang, Y.; Zhang, L.; Yin, J.; Hu, P.; Ma, E. Polarized Spectroscopic Properties of Ho^{3+}-Doped LuLiF$_4$ Single Crystal for 2 µm and 2.9 µm Lasers. *Opt. Mater.* **2011**, *33*, 1610–1615. [CrossRef]
34. Toncelli, A.; Tonelli, M.; Cassanho, A.; Jenssen, H. Spectroscopy and Dynamic Measurements of a Tm, Dy: BaY$_2$F$_8$ Crystal. *J. Lumin.* **1999**, *82*, 291–298. [CrossRef]
35. Cornacchia, F.; Toncelli, A.; Tonelli, M. 2-µm Lasers with Fluoride Crystals: Research and Development. *Prog. Quantum Electron.* **2009**, *33*, 61–109. [CrossRef]
36. Zhang, P.; Yin, J.; Zhang, B.; Zhang, L.; Hong, J.; He, J.; Hang, Y. Intense 2.8 µm Emission of Ho^{3+} Doped PbF$_2$ Single Crystal. *Opt. Lett.* **2014**, *39*, 3942–3945. [CrossRef] [PubMed]

37. Huang, X.; Wang, Y.; Zhang, P.; Su, Z.; Xu, J.; Xin, K.; Hang, Y.; Zhu, S.; Yin, H.; Li, Z. Efficiently Strengthen and Broaden 3 µm Fluorescence in PbF$_2$ Crystal by Er^{3+}/Ho^{3+} as Co-Luminescence Centers and Pr^{3+} Deactivation. *J. Alloy. Compd.* **2019**, *811*, 152027. [CrossRef]
38. Kresse, G.; Hafner, J. Ab Initio Molecular Dynamics for Liquid Metals. *Phys. Rev. B* **1993**, *47*, 558. [CrossRef] [PubMed]
39. Kresse, G.; Furthmüller, J. Efficiency of Ab-Initio Total Energy Calculations for Metals and Semiconductors Using a Plane-Wave Basis Set. *Comp. Mater. Sci.* **1996**, *6*, 15–50. [CrossRef]
40. Blöchl, P.E. Projector Augmented-Wave Method. *Phys. Rev. B* **1994**, *50*, 17953. [CrossRef] [PubMed]
41. Perdew, J.P.; Burke, K.; Ernzerhof, M. Generalized Gradient Approximation Made Simple. *Phys. Rev. Lett.* **1996**, *77*, 3865. [CrossRef]
42. Ma, F.; Su, F.; Zhou, R.; Ou, Y.; Xie, L.; Liu, C.; Jiang, D.; Zhang, Z.; Wu, Q.; Su, L. The Defect Aggregation of Re^{3+} (Re = Y, La ~ Lu) in MF$_2$ (M = Ca, Sr, Ba) Fluorites. *Mater. Res. Bull.* **2020**, *125*, 110788. [CrossRef]
43. Aull, B.; Jenssen, H. Vibronic Interactions in Nd: YAG Resulting in Nonreciprocity of Absorption and Stimulated Emission Cross Sections. *IEEE J. Quantum Electron.* **1982**, *18*, 925–930. [CrossRef]
44. Zhang, W.; Wang, Y.; Li, J.F.; Zhu, Z.J.; You, Z.Y.; Tu, C.Y. Spectroscopic Properties and Rate Equation Model of Er Doped BaLaGa$_3$O$_7$ Crystals. *Mater. Res. Bull.* **2018**, *106*, 282–287. [CrossRef]
45. Liu, Y.; Wang, Y.; You, Z.; Li, J.; Zhu, Z.; Tu, C. Growth, Structure and Spectroscopic Properties of Melilite Er: CaLaGa$_3$O$_7$ Crystal for Use in Mid-Infrared Laser. *J. Alloy. Compd.* **2017**, *706*, 387–394. [CrossRef]
46. Zhang, J.Y.; Han, S.J.; Liu, L.T.; Yao, Q.; Dong, W.M.; Li, J. Crystal Structure and Judd–Ofelt Analysis of Er^{3+} Doped LuAl$_3$(BO$_3$)$_4$ Crystal. *Chin. Phys. Lett.* **2018**, *35*, 096101. [CrossRef]
47. Hou, W.; Xu, Z.; Zhao, H.; Xue, Y.; Wang, Q.; Xu, X.; Xu, J. Spectroscopic Analysis of Er: Y$_2$O$_3$ Crystal at 2.7 µm Mid-IR Laser. *Opt. Mater.* **2020**, *107*, 110017. [CrossRef]

Article

A Study of the Characteristics of Plasma Generated by Infrared Pulse Laser-Induced Fused Silica

Lixue Wang, Xudong Sun, Congrui Geng, Zequn Zhang and Jixing Cai *

Jilin Key Laboratory of Solid-State Laser Technology and Application, Changchun University of Science and Technology, Changchun 130022, China; 17767764298@163.com (L.W.); masterpiecexd@outlook.com (X.S.); 18203227200@163.com (C.G.); Lugenze@outlook.com (Z.Z.)
* Correspondence: welcome8585@163.com; Tel.: +86-0431-85582187

Abstract: When high energy infrared laser pulses are incident on fused silica, the surface of the fused silica is damaged and a laser-induced plasma is produced. Based on the theory of fluid mechanics and gas dynamics, a two-dimensional axisymmetric gas dynamic model was established to simulate the plasma generation process of fused silica induced by a millisecond pulse laser. The results show that the temperature of the central region irradiated by the laser is the highest, and the plasma is first produced in this region. When the laser energy density is 1.0×10^4 J/cm^2 and the pulse width is 0.2 ms, the maximum expansion velocity of the laser-induced plasma is 17.7 m/s. Under the same experimental conditions, the results of the simulation and experiment are in good agreement. With an increase in pulse width, the plasma expansion rate gradually decreases.

Keywords: infrared pulse laser; fused silica; plasma; interaction between laser and matter; numerical simulation; plasma expansion velocity

Citation: Wang, L.; Sun, X.; Geng, C.; Zhang, Z.; Cai, J. A Study of the Characteristics of Plasma Generated by Infrared Pulse Laser-Induced Fused Silica. *Crystals* **2021**, *11*, 1009. https://doi.org/10.3390/cryst11081009

Academic Editors: Xiaoming Duan, Renqin Dou, Linjun Li and Xiaotao Yang

Received: 2 August 2021
Accepted: 18 August 2021
Published: 23 August 2021

Publisher's Note: MDPI stays neutral with regard to jurisdictional claims in published maps and institutional affiliations.

Copyright: © 2021 by the authors. Licensee MDPI, Basel, Switzerland. This article is an open access article distributed under the terms and conditions of the Creative Commons Attribution (CC BY) license (https://creativecommons.org/licenses/by/4.0/).

1. Introduction

With the popularization and application of a strong laser in various fields [1–4], optical elements as indispensable components of the laser system have received extensive attention. Because fused silica material has good light transmittance [5], it is widely used as an optical element in high power laser systems [6].

Fused silica material is one of the weakest components in a high power laser system. Its damage resistance directly affects the output performance of the laser. Even if there are subtle flaws, the performance of the laser will decrease. Therefore, it is of great significance to study the damage process of the interaction between the laser and fused silica [7,8]. R.W. Hopper et al. [9] established a theoretical model of laser damage induced by defects and studied the thermal stress damage caused by impurity defects in the target. F. Bolmeau et al. [10,11] used numerical simulation to study the laser energy deposition process caused by defects in fused silica.

The process of the interaction between the laser and fused silica is not only limited to material damage but also causes a series of processes such as phase transition, ablation, splashing, and plasma expansion. As a followup phenomenon of fused silica damage, plasma expansion has a high research value. However, the research on millisecond pulse laser-induced plasma generation of fused silica is quite scarce. Therefore, it is necessary to carry out related research on the plasma generation of fused silica induced by an infrared millisecond pulse laser.

As a high efficiency and commonly used numerical calculation method, the finite element analysis method is widely used in the treatment of various complex physical problems. M. Courtois et al. [12] used the finite element analysis method to numerically simulate and analyze the coupled heat transfer, fluid heat transfer, and thermal–mechanical coupling processes in laser welding. In this paper, the finite element method is used to simulate the plasma generation process of fused silica induced by the infrared millisecond pulse laser; the results of the simulation and experiment are in good agreement, and the results have good convergence.

2. Materials and Methods

In the process of laser heating the target, the state of the material changed from solid to liquid and then changed to a gaseous state. The laser interacted with the ejected gaseous material, the atoms were ionized, the plasma was formed, and ablation craters appeared in the laser-irradiated target area. The condensed target material spontaneously splashed into the action of atmosphere under the high pressure generated by the high temperature, which was the physical process of plasma expansion.

In the simulation process, considering plasma generation and wave propagation at geometric junctions, we used free triangular meshes to divide the computational domain. We refined the grids in the handover area and the upper area. The maximum cell size of the grid in the dense area was 0.02 mm, the smallest cell was 1.5×10^{-4} mm, the maximum cell growth rate was 1.08, and the curvature factor was 0.25. The time step used was the range of 0, 0.001, and 2, and the unit was millisecond.

Figure 1 is a schematic diagram of the numerical simulation of the laser-induced plasma. The initial plasma was generated on the surface of the target and propagated against the direction of the laser under the action of the laser.

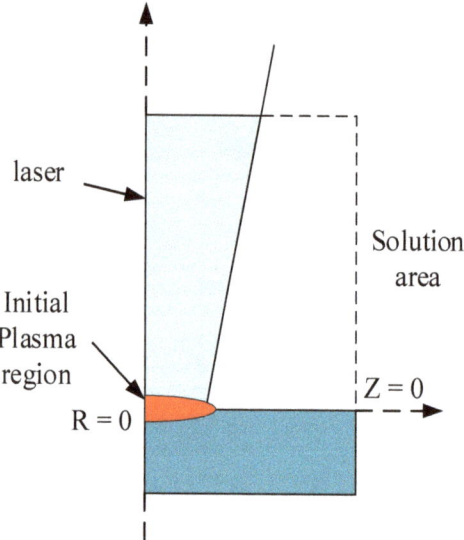

Figure 1. Schematic diagram of the initial plasma generation area.

The whole physical process of plasma generation and expansion satisfies conservation of mass, conservation of momentum, and conservation of energy. In the numerical simulation, the parameters of plasma changed with temperature T (where the units of T are K) are shown in Table 1.

The following needed to be clear during the simulation: (1) The propagation velocity of the plasma is much lower than the speed of sound, which can be regarded as subsonic laminar flow. (2) Plasma generation and expansion during a small span of time without heat exchange with the outside and inside is a local thermal equilibrium.

Table 1. Material parameters required for numerical simulation.

Feature Parameter Name	Symbol	Numerical Value
Plasma Density	$\rho/(\text{g}\cdot\text{cm}^{-3})$	$3.49/T \times 10^{-6}$
Plasma Thermal conductivity	$k/(\text{Wm}^{-1}\text{K}^{-1})$	$-0.002 + 1.5 \times 10^{-4} \times T - 7.9 \times 10^{-8} \times T^2 + 4.12 \times 10^{-11} \times T^3 - 7.44 \times 10^{-15} \times T^4$
Plasma Heat capacity	$C/(\text{J}/(\text{kg}\cdot\text{K}))$	$1047.27 + 9.45 \times 10^{-4} \times T^2 - 6.02 \times 10^{-7} \times T^3 + 1.29 \times 10^{-10} \times T^4$
Plasma Viscosity coefficient	$\eta/(\text{Pa}\cdot\text{S})$	$-8.38 \times 10^{-7} + 8.36 \times 10^{-8} \times T - 7.69 \times 10^{-11} \times T^2 + 4.64 \times 10^{-14} \times T^3 - 1.07 \times 10^{-17} \times T^4$
Melting point	T_m/K	1730
Boiling point	T_v/K	2503
Molar mass	$M_{SiO_2}/(\text{g}/\text{mol})$	60

Throughout the physical process, the total mass of matter remained unchanged, which satisfied the law of conservation of mass:

$$\nabla \cdot (\rho u) = 0 \quad (1)$$

In the formula: u is the fluid velocity, ρ is the density.

Momentum conservation equation:

$$\rho\left(\frac{\partial u}{\partial t} + u \cdot (\nabla \cdot u)\right) = \nabla \cdot \left(-p + \mu\left(\nabla u + (\nabla u)^T\right)\right) \quad (2)$$

In the formula: p is the pressure, μ is the material viscosity.

Energy conservation equation:

$$\rho C_P \frac{\partial T}{\partial t} + \rho u C_P \nabla T = \nabla(k\nabla T) + (Q_{Laser} - Q_{Loss}) \quad (3)$$

In the formula: C_P is the specific heat capacity, K is the thermal conductivity, Q_{Laser} is the laser heat source, and Q_{Loss} is the heat source that maintains the plasma expansion loss.

In the energy conservation equation, Q_{Laser} is one of the source terms of the laser heat source; it is simplified to the Gaussian surface heat source distribution that varies with radius and temperature:

$$Q_L = A(T)\frac{2I_0}{\pi r_0^2}\exp\left(\frac{-2r^2}{r_0^2}\right)\tau(t) \quad (4)$$

where I_0 is the power density, r_0 is the laser spot radius, and $A(T)$ is the dynamic absorption rate.

$\tau(t)$ is the time distribution function of the laser pulse width; the expression is as follows:

$$\tau(t) = \begin{cases} 1, t \leq \tau_P \\ 0, t \geq \tau_P \end{cases} \quad (5)$$

τ_P is the laser pulse width.

The vaporization and expansion process included the vapor evaporation process of the material leaving the target surface and the process of spraying outwards under continued heating. The mass loss of the surface layer of the fused silica material during the outward spraying led to the further formation of molten pits and continued to induce plasma. The quantity that affects the quality loss is the mass mobility [12]:

$$m_0 = \sqrt{\frac{M_{SiO_2}}{2\pi k_B}} \frac{P_{sat}(T)}{\sqrt{T}} \beta \tag{6}$$

where β is the diffusion coefficient, the value is assumed to be 1 at the beginning of the evaporation, and the value is 0 when it is stabilized, k_B is Boltzmann's constant, $P_{sat}(T)$ is the saturated vapor pressure of the target steam [12,13], and the expression is:

$$P_{sat}(T) = P_{amb} \exp\left(\frac{L_v M_{SiO_2}}{k_B T}\left(\frac{T}{T_v} - 1\right)\right) \tag{7}$$

where P_{amb} is the atmospheric pressure, L_v is the latent heat of evaporation, T_v is the evaporation temperature, and M_{SiO_2} is the molar mass of the fused silica.

In Equation (3), the expression of Q_{Loss} is:

$$Q_{Loss} = L_V v_r \rho_l + k\frac{\partial T}{\partial n} \tag{8}$$

ρ_l is the density of the material after melting, L_V is the latent heat of vaporization, v_r is the expansion speed of the plasma, and k is the energy loss rate in the system.

Computational domain boundary conditions:

$$-k\frac{\partial T(r,z,t)}{\partial z}\bigg|_{z=h, z=b} = 0 \tag{9}$$

3. Principle of the Experiment

The experimental measurement principle diagram of the plasma expansion of fused silica induced by the millisecond pulse laser is shown in Figure 2.

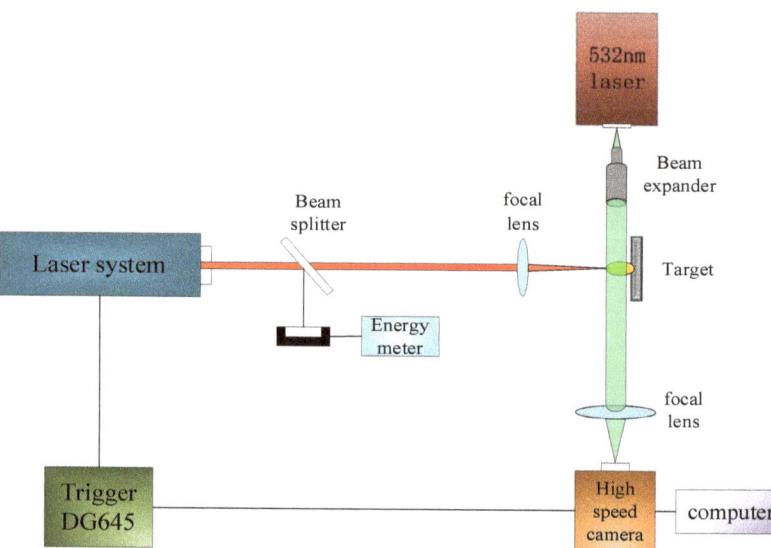

Figure 2. Experimental setup.

In order to analyze the expansion law of the plasma generated by laser-induced fused silica, the high-time-resolution optical shadow method was used to monitor the evolution of the laser-induced plasma flow field. The time delay between the high-speed camera and the laser system was adjusted by using the DG645 digital delay generator. The high-speed camera received the image. By analyzing the optical shadow images under different time series conditions, the expansion and propagation process of the laser-induced plasma was obtained.

4. Results and Discussion

4.1. Analysis of the Plasma Temperature Field Induced by a Pulsed Laser in Fused Silica

Under millisecond pulse laser irradiation, the laser energy densities used to simulate the plasma generated by laser irradiation of fused silica were 8.0×10^3 J/cm^2, the pulse width was 0.2 ms, and the irradiation radius was 0.25 mm.

The temperature field results of the numerical simulation of the plasma generated by the millisecond pulse laser induced fused silica were as follows:

When the laser with an energy density of 8.0×10^3 J/cm^2 irradiated the surface of the fused silica material, the main form of plasma generated in the laser irradiation area was thermally induced plasma. Due to the laser heating, the surface of the fused silica material melted, vaporized, and splashed, forming a numerical simulation of the low-temperature thermally induced plasma temperature field as shown in the figure.

Figure 3 shows the temperature field distribution of the laser-induced plasma at different times of the plasma. At the initial stage of laser action 0–0.06 ms, the temperature of the irradiated area rose rapidly and plasma began to be generated; the temperature gradually stabilized within the time range of 0.10–0.15 ms, the plasma slowly expanded outward, and thermally induced plasma was generated. In the time range of 0.25–0.30 ms, since the internal energy of the thermally induced plasma was converted into kinetic energy, the plasma expanded outwards and the temperature decreased accordingly.

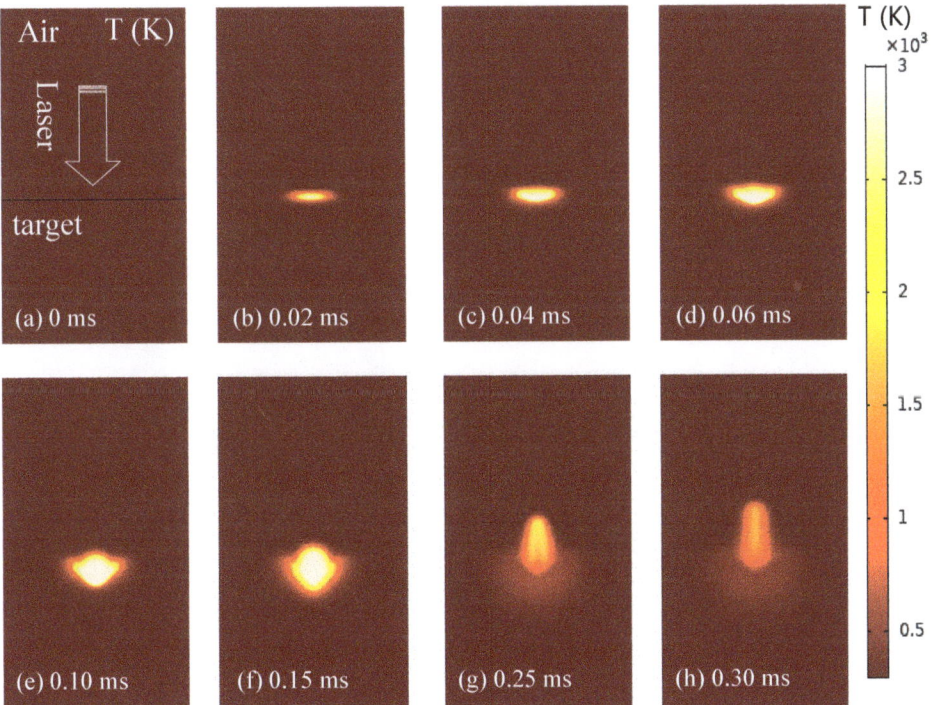

Figure 3. Time-resolved simulation results of the plasma temperature field, subfigures (**a**–**h**) are the distribution diagrams of the plasma temperature field at different moments.

Figure 4 is the curve of radial temperature change with time when the laser energy density was 8.0×10^3 J/cm^2 in the period of 0.005–0.030 ms. From Figure 4, it can be seen that the temperature change in the center area of the laser irradiation on the surface of the target was much greater than that of other areas. This is because the pulse laser has a Gaussian distribution, and the laser energy is concentrated in the center, which makes the highest temperature rise in the irradiated center area the fastest. Figure 5 is the curve of radial temperature change with time when the laser energy density was 8.0×10^3 J/cm^2 in the period of 0.035–0.055 ms. Compared with Figure 4, it can be seen that the temperature in the central area of laser irradiation was stable at 0.030 ms, and plasma began to be generated. At this time, since the action time of the pulsed laser did not stop, the generation area of the plasma began to expand.

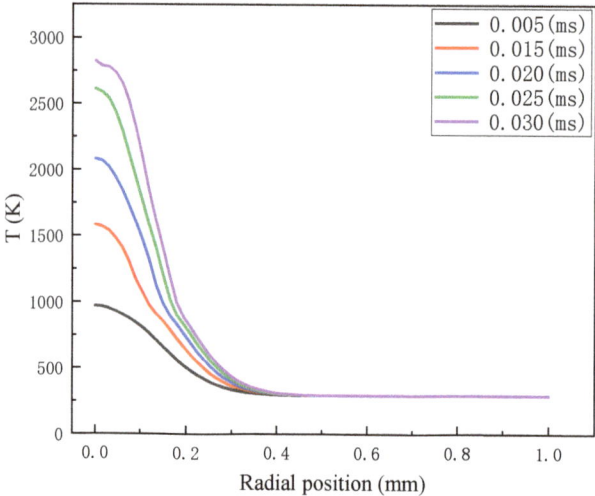

Figure 4. Simulation results of the radial temperature changing with time when the laser energy density was 8.0×10^3 J/cm^2 in the time period of 0.005–0.030 ms.

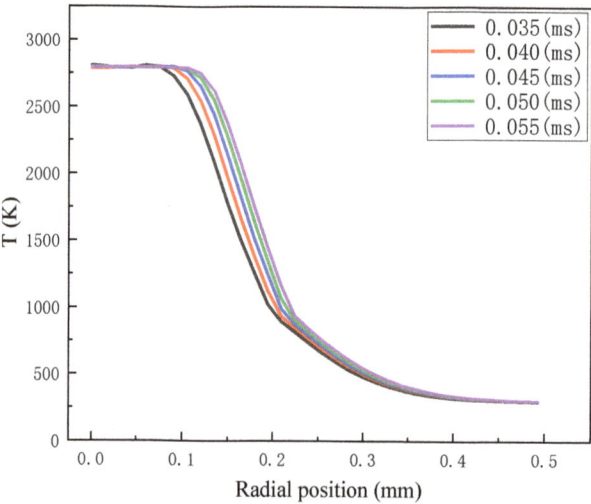

Figure 5. Simulation results of the radial temperature change with time when the laser energy density was 8.0×10^3 J/cm^2 in the time period of 0.035–0.055 ms.

4.2. Analysis of the Plasma Flow Field in Fused Silica Induced by a Pulsed Laser

Figure 6 shows the distribution of the plasma flow field caused by the laser with a laser energy density of 1.0×10^4 J/cm^2 and a pulse width of 0.2 ms acting on the fused silica.

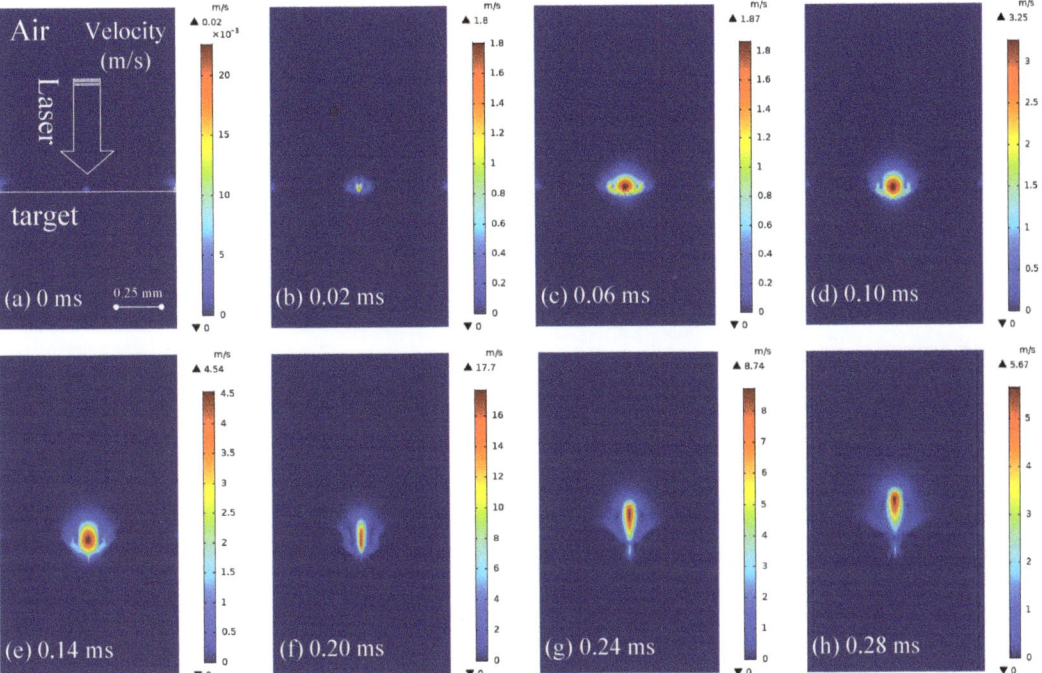

Figure 6. Simulation results graph of the plasma velocity evolution with time under 1.0×10^4 J/cm^2 laser energy density, subfigures (a–h) are the distribution diagrams of the plasma flow field at different moments.

Figure 6 shows the flow field distribution of the plasma at different times. In the initial stage of pulse laser action, the velocity of the flow field in the action area increased gently (0–0.14 ms). In the time range of 0.14–0.20 ms, the plasma flow field velocity increased rapidly and reached maximum velocity at 0.2 ms, with a value of 17.7 m/s. The velocity of the plasma flow field decreased gradually between 0.24 ms and 0.28 ms. This was because the laser action stopped, and the laser irradiation did not continue to provide energy; the plasma could only rely on its own internal energy to be converted into kinetic energy to continue to maintain its outward expansion.

The plasma expansion velocity as a function of time t was found using

$$v(t) = \frac{L(t+t) - L(t)}{t} \tag{10}$$

where the expanding plasma front is measured to reach distance $L(t)$ at time t, and t is a small interval of time typically; the unit of t we used here is μs [14].

Figure 7A shows the axial velocity distribution of plasma generated by the millisecond laser-induced fused silica with a pulse width of 0.2 ms; the dotted line is the experimental results, the solid line is the simulation results. Within 0–0.2 ms, the laser induced plasma continuously absorbed laser energy and expanded outward rapidly. Figure 7B shows that the axial plasma expansion rate was much greater than the radial expansion rate. The reason for this phenomenon is that the plasma expansion speed is determined by the laser energy. When the laser was incident perpendicular to the target surface, the laser energy distributed in the axial direction was much greater than the laser energy in the radial direction. The axial position made it easier for the plasma to move faster than in the radial position. Therefore, there was a large gap between the expansion speed of the axial plasma and the expansion speed of the radial plasma.

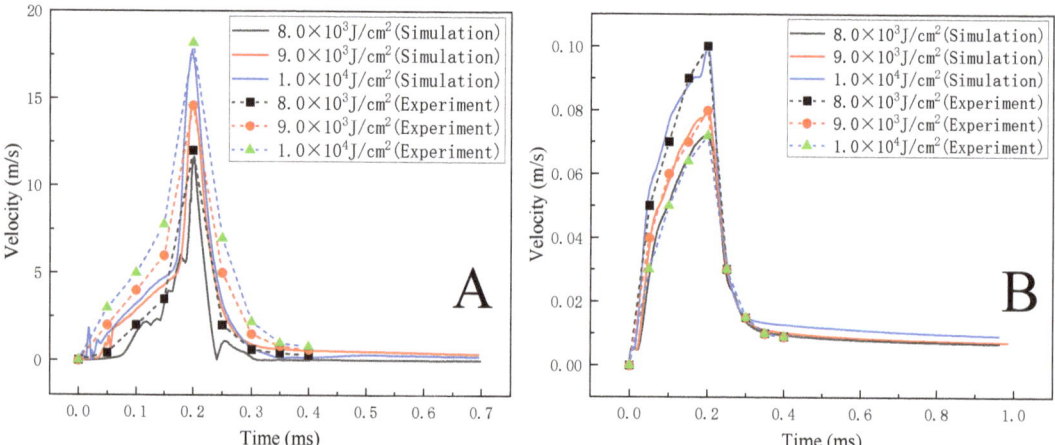

Figure 7. The evolution curve of the maximum plasma velocity field with time under different laser energies; (**A**) Axial plasma flow field velocity curve; (**B**) Radial plasma flow field velocity curve.

Figure 8 shows the evolution process of the plasma expansion based on the shadow method using a high-speed camera with a pulse width of 0.2 ms under the same laser energy density (which was 50 µs/frame). Through calculation, we obtained the maximum expansion velocity of the combustion wave in this process, about 18.2 m/s. The observed experimental phenomena were in good agreement with our calculation results.

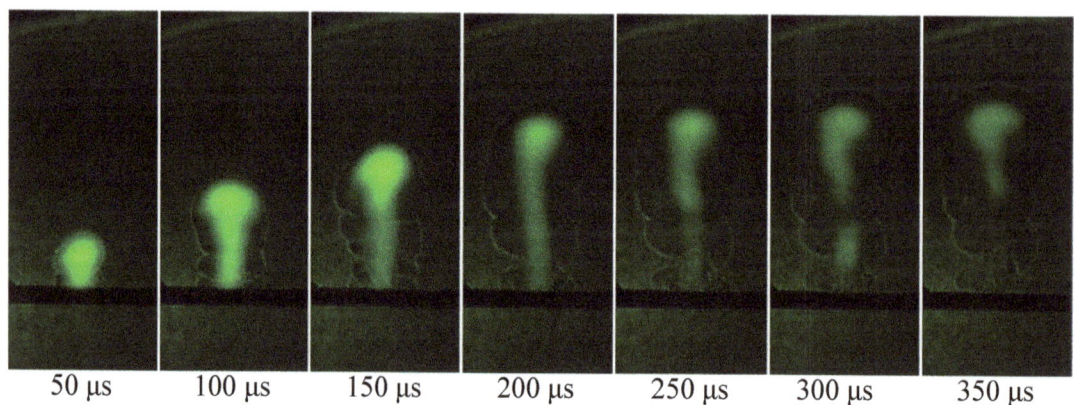

Figure 8. Plasma expansion process image.

Figure 9 shows the relationship between the plasma velocity field and the pulse width at three different energy densities of 8.0×10^3 J/cm^2, 9.0×10^3 J/cm^2, and 1.0×10^4 J/cm^2 (Figure 9A–D), respectively, corresponding to the pulse widths of 0.2 ms, 0.5 ms, 0.8 ms, and 1.0 ms.

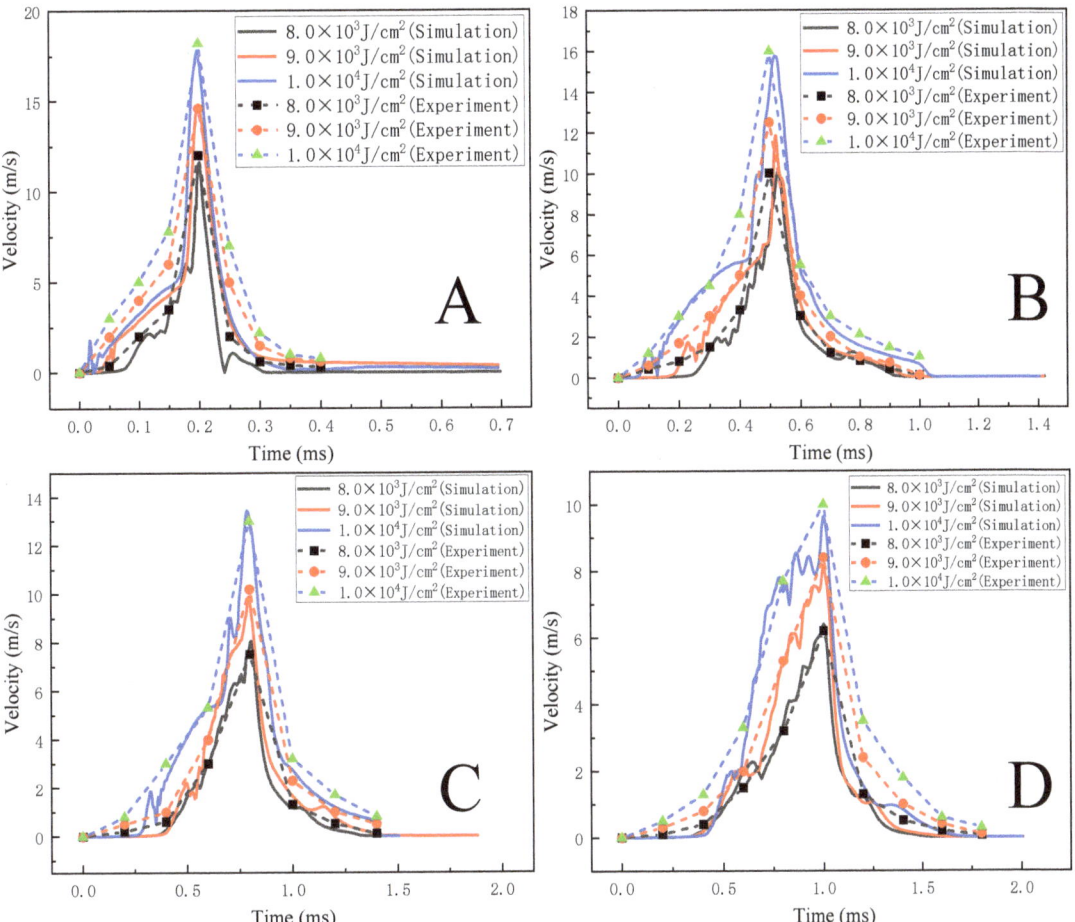

Figure 9. Distribution of the plasma flow field under different pulse widths; (**A**) The evolution of the plasma flow field with time when the pulse width was 0.2 ms; (**B**) The evolution of the plasma flow field with time when the pulse width was 0.5 ms; (**C**) The evolution of the plasma flow field with time when the pulse width was 0.8 ms; (**D**) The evolution of the plasma flow field with time when the pulse width was 1 ms.

Figure 10 shows the relationship between the plasma expansion velocity field and the laser pulse width under three different laser energy densities. From Figure 10, it can be seen that when the laser energy density was 8.0×10^3 J/cm^2, the maximum speeds corresponding to 0.2 ms, 0.5 ms, 0.8 ms, and 1.0 ms were 11.545 m/s, 10.169 m/s, 8.0608 m/s, and 6.3985 m/s. When the energy density was 9.0×10^3 J/cm^2, the maximum speeds corresponding to the simulation results were 14.76 m/s, 11.848 m/s, 9.8884 m/s, and 8.159 m/s, respectively. When the energy density was 1.0×10^4 J/cm^2, the corresponding maximum speeds were 17.7 m/s, 15.769 m/s, 13.453 m/s, and 10.046 m/s, respectively.

Comparing the plasma flow field velocity in Figures 8 and 9, it can be seen that the time for the plasma flow field velocity to reach the maximum value increased with the increase in the pulse width, and the maximum velocity decreased with the increase in the pulse width. After the laser ended, the plasma flow field velocity rapidly decreased to zero.

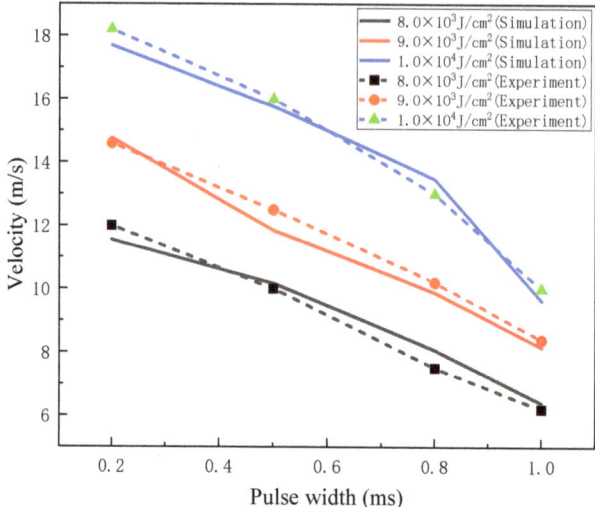

Figure 10. The relationship between the velocity field and pulse width under different laser energy densities.

5. Conclusions

This paper considered the laser energy transmission processes including thermal radiation, heat conduction, and convection, and simulated the whole physical process of the millisecond pulse laser induced fused silica to generate plasma, and the simulation results have good convergence. The simulation of laser-induced fused silica plasma can easily provide some details, such as the initial plasma temperature, which is not easy to measure in the experiment. It can deepen people's understanding of the initial details of plasma generation and make the plasma generation process less abstract. The plasma simulation model still has some shortcomings, and people need to continue to improve it in future work, so that it can represent a more realistic physical process. The results were as follows:

(1) The temperature field mainly discussed the relationship of temperature distribution in the radial position over time. The simulation results showed that the temperature in the center region irradiated by the laser was the highest, and plasma was first generated in this region.

(2) In the velocity field, the plasma flow field under different laser energy densities and pulse widths were analyzed and studied, and the relationship between the plasma expansion law and the laser energy density and pulse width were obtained. The simulation results showed that there was a large gap between the axial plasma expansion speed and the radial plasma expansion speed; the plasma expansion speed increased with the increase in the laser energy density and decreased with the increase in the pulse width. The simulation results were in good agreement with the experimental results, and it also provided a theoretical and experimental basis for studying the laser-induced plasma process of fused silica materials.

Author Contributions: Conceptualization, L.W. and Z.Z.; methodology, C.G.; software, C.G.; validation, L.W., X.S. and Z.Z.; formal analysis, L.W.; investigation, L.W.; resources, L.W.; data curation, L.W.; writing—original draft preparation, L.W.; writing—review and editing, J.C.; visualization, L.W.; supervision, J.C.; project administration, J.C.; funding acquisition, J.C. All authors have read and agreed to the published version of the manuscript.

Funding: This research was funded by the National Natural Science Foundation of China, grant number 62005023.

Institutional Review Board Statement: Not applicable.

Informed Consent Statement: Not applicable.

Data Availability Statement: Not applicable.

Acknowledgments: Thanks to the Jilin Key Laboratory of Solid-State Laser Technology and Application, School of Science, Changchun University of Science and Technology.

Conflicts of Interest: The authors declare no conflict of interest.

References

1. Zhuang, S.; Kainuma, S.; Yang, M.; Haraguchi, M.; Asano, T. Characterizing corrosion properties of carbon steel affected by high-power laser cleaning. *Constr. Build. Mater.* **2021**, *274*, 122085. [CrossRef]
2. Schneider, F.; Wolf, N.; Petring, D. High Power Laser Cutting of Fiber Reinforced Thermoplastic Polymers with cw- and Pulsed Lasers. *Phys. Procedia* **2013**, *41*, 415–420. [CrossRef]
3. Ohnishi, T.; Kawahito, Y.; Mizutani, M.; Katayama, S. High-Power and High-Brightness Laser Butt Welding with Using Hot Wire for Thick High-Strength Steel Plate. *Q. J. Jpn. Weld. Soc.* **2011**, *29*, 41–47. [CrossRef]
4. Wu, J.; Zhang, H.; Li, Y.; Zhong, Q. Numerical analysis of mass distribution in side shielding gas in laser welding. *HanjiXuebao/Trans. China Weld. Inst.* **2018**, *39*, 39–43.
5. Kotz, F.; Quick, A.S.; Risch, P.; Martin, T.; Hoose, T.; Thiel, M.; Helmer, D.; Rapp, B.E. Two-Photon Polymerization of Nanocomposites for the Fabrication of Transparent Fused Silica Glass Microstructures. *Adv. Mater.* **2021**, *33*, 2006341. [CrossRef] [PubMed]
6. Yoshida, H.; Fujita, H.; Nakatsuka, M.; Yoshida, K. High Resistant Phase-conjugated Stimulated Brillouin Scattering Mirror Using Fused-silica Glass for Nd:YAG Laser System. *Jpn. J. Appl. Phys.* **1999**, *38*, L521–L523. [CrossRef]
7. Kozlowski, M.R.; Thomas, I.M.; Campbell, J.H.; Rainer, F. High-power optical coatings for a megajoule-class ICF laser. In Proceedings of the SPIE, Berlin, Germany, 14–19 September 1992; pp. 105–119.
8. Wang, B.; Dai, G.; Zhang, H.; Ni, X.; Shen, Z.; Lu, J. Damage performance of TiO_2/SiO_2 thin film components induced by a long-pulsed laser. *Appl. Surf. Sci.* **2011**, *257*, 9977–9981. [CrossRef]
9. Hopper, R.W.; Ublmann, D.R. Mechanism of inclusion damage in laser glass. *J. Appl. Phys.* **1970**, *41*, 4023–4037. [CrossRef]
10. Bonneau, F.; Combis, P.; Rullier, J.L.; Vierne, J.; Pellin, M.; Savina, M.; Broyer, M.; Cottancin, E.; Tuaillon, J.; Pellarin, M.; et al. Study of UV laser interaction with gold nanoparticles embedded in silica. *Appl. Phys. B* **2002**, *75*, 803–815. [CrossRef]
11. Bonneau, F.; Combis, P.; Rullier, J.L.; Vierne, J.; Bertussi, B.; Commandre, M.; Gallais, L.; Natoli, J.Y.; Bertron, I.; Malaise, F.; et al. Numerical simulations for description of UV laser interaction with gold nanoparticles embedded in silica. *Appl. Phys. B* **2004**, *78*, 447–552. [CrossRef]
12. Courtois, M.; Carin, M.; Le Masson, P.; Gaied, S.; Balabane, M. A new approach to compute multi-reflections of laser beam in a keyhole for heat transfer and fluid flow modelling in laser welding. *J. Phys. D Appl. Phys.* **2013**, *46*, 505305. [CrossRef]
13. Ranjbar, O.A.; Lin, Z.; Volkov, A.N. One-dimensional kinetic simulations of plume expansion induced by multi-pulse laser irradiation in the burst mode at 266 nm wavelength. *Vacuum* **2018**, *157*, 361–375. [CrossRef]
14. Ji-Xing, C.; Ming, G.; Xu, Q.; He, L.; Guang-Yong, J. Gas dynamics and combustion wave expanding velocity of laser induced plasma. *Acta Phys. Sin.* **2017**, *66*, 094202. [CrossRef]

Article

Investigation of Mid-Infrared Broadband Second-Harmonic Generation in Non-Oxide Nonlinear Optic Crystals

Ilhwan Kim, Donghwa Lee and Kwang Jo Lee *

Department of Applied Physics, Institute of Natural Science, Kyung Hee University, Yongin-si 17104, Korea; dlfghks383@gmail.com (I.K.); fairytale095@gmail.com (D.L.)
* Correspondence: kjlee88@khu.ac.kr

Abstract: The mid-infrared (mid-IR) continuum generation based on broadband second harmonic generation (SHG) (or difference frequency generation) is of great interest in a wide range of applications such as free space communications, environmental monitoring, thermal imaging, high-sensitivity metrology, gas sensing, and molecular fingerprint spectroscopy. The second-order nonlinear optic (NLO) crystals have been spotlighted as a material platform for converting the wavelengths of existing lasers into the mid-IR spectral region or for realizing tunable lasers. In particular, the spectral coverage could be extended to ~19 μm with non-oxide NLO crystals. In this paper, we theoretically and numerically investigated the broadband SHG properties of non-oxide mid-IR crystals in three categories: chalcopyrite semiconductors, defect chalcopyrite, and orthorhombic ternary chalcogenides. The technique is based on group velocity matching between interacting waves in addition to birefringent phase matching. We will describe broadband SHG characteristics in terms of beam propagation directions, spectral positions of resonance, effective nonlinearities, spatial walk-offs between interacting beams, and spectral bandwidths. The results will show that the spectral bandwidths of the fundamental wave allowed for broadband SHG to reach several hundreds of nm. The corresponding SH spectral range spans from 1758.58 to 4737.18 nm in the non-oxide crystals considered in this study. Such broadband SHG using short pulse trains can potentially be applied to frequency up-conversion imaging in the mid-IR region, in information transmission, and in nonlinear optical signal processing.

Keywords: mid-infrared photonics; chalcopyrite semiconductors; orthorhombic ternary chalcogenides; harmonic generation; continuum generation; group velocity matching

1. Introduction

The field of mid-infrared (IR) photonics is growing rapidly due to increasing demand for applications such as free space communications, remote sensing, environmental monitoring, thermal imaging, defense, IR countermeasure, medicine, gas sensing, and molecular fingerprint spectroscopy [1–3]. Nonlinear optic (NLO) crystals—particularly with second-order nonlinearities—have been spotlighted as a material platform for converting the wavelengths of existing lasers into the mid-IR spectral region or for realizing tunable lasers [4,5]. The mid-IR spectral regions up to ~4 μm could be readily accessible with the oxide NLO crystals such as lithium niobate ($LiNbO_3$), lithium tantalate ($LiTaO_3$), and potassium titanyl phosphate ($KTiOPO_4$) [6–9]. A high peak-power mid-IR optical amplifier using a potassium titanyl arsenate ($KTiOAsO_4$) is also reported in [10]. These oxide crystals have been mainly used in the near-IR region but are still transparent within 4 μm. However, the spectral coverage of oxide NLO crystals is limited to less than 5 μm due to multi-phonon absorption [4]. The upper spectral limit of photons generated via parametric generation can be extended to ~19 μm using non-oxide crystals such as chalcopyrite semiconductors, orthorhombic ternary chalcogenides, and orientation-patterned (OP) semiconductors (e.g., OP-GaAs, OP-GaP, OP-ZnSe, and OP-GaN) [4,5,11,12]. The OP semiconductors with periodic inversion of crystalline orientation use quasi-phase matching (QPM) for NLO interactions, whereas chalcopyrite semiconductors and orthorhombic ternary chalcogenides generally utilize birefringent phase matching (BPM). Extensive experimental studies of mid-IR laser sources using chalcopyrite semiconductors (e.g., silver thiogallate (AgGaS$_2$, AGS), silver gallium selenide (AgGaSe$_2$, AGSe), cadmium silicon phosphide (CdSiP$_2$, CSP), and zinc germanium phosphide (ZnGeP$_2$, ZGP)), defect chalcopyrite crystals (e.g., mercury thiogallate (HgGa$_2$S$_4$, HGS) and cadmium selenide (CdSe)), and orthorhombic ternary chalcogenides (e.g., lithium thioindate (LiInS$_2$, LIS), lithium thiogallate (LiGaS$_2$, LGS), and lithium gallium selenide (LiGaSe$_2$, LGSe)) have been reported continuously in recent years [13–26].

Continuum generation based on broadband second harmonic generation (SHG) (or difference frequency generation) is of great interest in a wide range of applications. Considering the bandwidth of the continuum is inversely proportional to the group velocity (GV) mismatch between the interacting optical waves, broadband parametric generation is possible through GV matching [12]. Such broadband SHG using short pulse trains can potentially be applied to frequency up-conversion imaging in the mid-IR region, in information transmission using optical pulse signals, and in nonlinear optical signal processing [27–29]. For example, a wave packet transmitted through a free-space communication system can be decomposed into individual pulse signals by Fourier analysis and then each optical signal containing information is modulated and controlled for by another clock pulse train [27]. Recently, the potential to create such a continuum using ultrashort-pulse lasers has been estimated, especially in oxide crystals [12,30]. However, for non-oxide NLO crystals, broadband parametric generation has not yet been intensively studied. In this paper, we theoretically and numerically investigate broadband SHGs in three categories of commercially available non-oxide NLO crystals: (1) chalcopyrite semiconductors exhibiting uniaxial birefringence (e.g., cadmium germanium arsenide (CdGeAs$_2$, CGA), AGS, AGSe, CSP, and ZGP); (2) defect chalcopyrite showing uniaxial birefringence (e.g., thallium arsenic selenide (Tl$_3$AsSe$_3$, TASe), gallium selenide (GaSe), CdSe, and HGS); and (3) orthorhombic ternary chalcogenides showing biaxial birefringence (e.g., lithium indium selenide (LiInSe$_2$, LISe), LIS, LGS, and LGSe). The technique is based on GV matching between interacting waves in addition to BPM. In this simultaneous BPM–GV matching approach, the GV mismatch (GVM) between the interacting waves is always zero at the resonances within a specific spectral range that satisfies both the BPM and zero GVM. For uniaxial crystals, the wavelengths satisfying the simultaneous BPM–GV matching scheme correspond to the "magic" wavelengths in oxide crystals as reported in [12,30]. For biaxial crystals, these magic wavelengths extend into specific magic spectral regions, as will be discussed later. The advantage of this scheme is that the acceptable bandwidth of the fundamental (F) wave becomes very broad. This broadband spectrum has potential applications in multi-channel nonlinear optic signal processing. A broader input bandwidth also means that a train of pulses with narrower temporal widths can be used as the F-wave because the temporal width of a pulse has a Fourier transform relationship with its spectral width. Another critical advantage of the simultaneous BPM–GV matching scheme is that it allows for the use of long crystal lengths without considering the temporal walk-off between the interacting waves. The NLO efficiency is proportional to the square of the crystal length. Of course, there is still a limit on the crystal length due to the spatial walk-off between the interacting waves but this is significantly smaller than the limit due to the GVM, as will be described in Section 3. For each kind of non-oxide crystal, Type I and Type II NLO interactions will be considered, in which the polarizations of two fundamental photons to produce a second harmonic (SH) photon are either parallel to each other (for Type I) or perpendicular to each other (for Type II). We will describe the broadband SHG characteristics of each crystal in terms of beam propagation directions, spectral positions of resonance, effective nonlinearities, spatial walk-offs between interacting beams, and spectral bandwidths. The results will show that the spectral bandwidths of the F-wave allowed for the broadband SHG to reach several hundreds of nm. The SH spectral range satisfying the simultaneous BPM–GV matchings span from 1758.58 to 4737.18 nm in the non-oxide crystals considered in this study. Finally, potential applications of such broadband light sources will be briefly described.

2. Materials and Theories

The non-oxide NLO crystals considered in this work are listed in Table 1. As can be seen from Table 1, chalcopyrite semiconductors and defective chalcopyrite exhibit either positive or negative uniaxial birefringence, whereas orthorhombic ternary chalcogenides have biaxial birefringence. As will be described later, the type of birefringence determines the angle tunability of the spectral position of NLO resonance. The point groups for each crystal determines the effective nonlinearity for the given direction of the input F-beam and are listed together in Table 1. The transparent range for each crystal is summarized in Table 1 along with references. In this section, we will describe the theoretical details of the BPM and GV matching properties of non-oxide NLO crystals, the effective nonlinearities, and the spatial walk-offs between the interacting beams.

Table 1. Non-oxide nonlinear optic (NLO) crystals considered for broadband mid-IR second-harmonic generation (SHG).

Crystals	Birefringence	Crystal System	Point Group	Transparency
AgGaS$_2$ (AGS)	Negative uniaxial	Tetragonal	$\bar{4}2m$	0.45–13 µm [31]
AgGaSe$_2$ (AGSe)				0.71–19 µm [32]
CdSiP$_2$ (CSP)				0.66–6.5 µm [33]
HgGa$_2$S$_4$ (HGS)			$\bar{4}$	0.55–11 µm [34]
Tl$_3$AsSe$_3$ (TASe)		Trigonal	$3m$	1.26–17 µm [35]
GaSe		Hexagonal	$\bar{6}2m$	0.65–18 µm [36]
CdGeAs$_2$ (CGA)	Positive uniaxial	Tetragonal	$\bar{4}2m$	2.4–18 µm [37]
ZnGeP$_2$ (ZGP)				0.74–12 µm [36]
CdSe		Hexagonal	$6mm$	0.8–20 µm [38]
LiInS$_2$ (LIS)	Negative biaxial	Orthorhombic	$mm2$	0.4 µm–12 µm [39]
LiInSe$_2$ (LISe)				0.5 µm–12 µm [40]
LiGaS$_2$ (LGS)				0.32 µm–11.6 µm [41]
LiGaSe$_2$ (LGSe)				0.37 µm–13.2 µm [41]

2.1. BPM and GV Matching for Broadband SHG

Figure 1 illustrates the polarization relationships of F and SH waves. H and V in Figure 1 represent horizontal and vertical polarization directions, respectively. For Type I, a pair of F photons with the same frequencies, ω, and polarization states produces an SH photon with a frequency, 2ω. In this case, the SH photon and F photons have polarization states perpendicular to each other (Figure 1a). For Type II, a pair of F photons with H and V polarization states, respectively, generates an SH photon (Figure 1b). In the experiment, a pair of photons with H and V polarization states were obtained from the input light polarized at 45° to the horizontal direction, as shown in Figure 1b.

In uniaxial crystals, the order of the refractive index (RI) magnitude is given by either $n_e < n_o$ (for negative uniaxial crystals: AGS, AGSe, CSP, HGS, TASe, and GaSe) or $n_e > n_o$ (for positive uniaxial crystals: CGA, ZGP, and CdSe), where n_o and n_e represent the RIs of the ordinary (o) and extraordinary (e) waves in the principal axes of an index ellipsoid [42]. Then, the collinear BPM conditions can be expressed as:

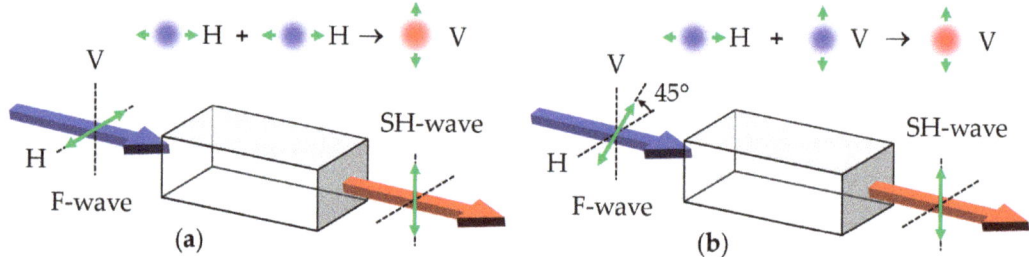

Figure 1. Schematic diagrams showing the polarization relationships of fundamental (F) and second harmonic (SH) waves for (**a**) Type I and (**b**) Type II. V and H represent vertical and horizontal polarization directions, respectively.

$$\Delta k = |k_e(2\omega,\theta) - 2k_o(\omega)| = 0 \text{ (for type I, negative uniaxial)}, \quad (1)$$

$$\Delta k = |k_e(2\omega,\theta) - k_o(\omega) - k_e(\omega,\theta)| = 0 \text{ (for type II, negative uniaxial)}, \quad (2)$$

$$\Delta k = |k_o(2\omega) - 2k_e(\omega,\theta)| = 0 \text{ (for type I, positive uniaxial)}, \quad (3)$$

$$\Delta k = |k_o(2\omega) - k_o(\omega) - k_e(\omega,\theta)| = 0 \text{ (for type II, positive uniaxial)}, \quad (4)$$

where each k represents the wave number of the interacting wave and is defined as $k_o(j\omega) = (j\omega/c)n_o(j\omega)$ or $k_e(j\omega,\theta) = (j\omega/c)n_e(j\omega,\theta)$. Here, j can be 1 or 2 and c denotes the speed of light in vacuum. For light propagating at the angle θ to the optic axis of uniaxial birefringence, the RI of e-wave, $n_e(\theta)$, can be derived as follows using the definition in [42]:

$$n_e(j\omega,\theta) = \frac{n_o(j\omega)n_e(j\omega)}{\sqrt{n_o^2(j\omega)\sin^2\theta + n_e^2(j\omega)\cos^2\theta}}. \quad (5)$$

The temporal walk-off between the interacting waves due to the difference in GV can be defined as the time delay (ΔT) per unit crystal length as follows:

$$\frac{\Delta T}{L} = \frac{\Delta n_g}{c}, \quad (6)$$

where L and Δn_g represent the crystal length and the group index difference between interacting photons, respectively. When $\Delta n_g = 0$ in Equation (6), GV matching is achieved, which can be simplified as:

$$n_e^{(g)}(2\omega,\theta) = n_o^{(g)}(\omega) \text{ (for type I, negative uniaxial)}, \quad (7)$$

$$2n_e^{(g)}(2\omega,\theta) = n_o^{(g)}(\omega) + n_e^{(g)}(\omega,\theta) \text{ (for type II, negative uniaxial)}, \quad (8)$$

$$n_o^{(g)}(2\omega) = n_e^{(g)}(\omega,\theta) \text{ (for type I, positive uniaxial), and} \quad (9)$$

$$2n_o^{(g)}(2\omega) = n_o^{(g)}(\omega) + n_e^{(g)}(\omega,\theta) \text{ (for type II, positive uniaxial)}, \quad (10)$$

where each superscript g in Equations (7)–(10) indicates the group index. Now, the broadband SHG in a negative uniaxial crystal is defined as Equations (1) and (7) (for Type I) or Equations (2) and (8) (for Type II), being satisfied simultaneously. For a positive uniaxial crystal, the broadband SHG condition is defined as Equations (3) and (9) (for Type I) or Equations (4) and (10) (for Type II). Table 2 summarizes the BPM and GV matching conditions for broadband SHG in uniaxial crystals. All equations in Table 2 are expressed as two-variable functions for θ and ω (or for the F-wavelength, λ_F). Solving the system of equations for BPM and GV matching yields a solution set of λ_F and θ, which determine the center wavelength of broadband SHG and the propagation direction of the F-beam.

Table 2. The BPM and GV matching conditions for broadband SHG in uniaxial crystals.

Type	Condition	Negative Uniaxial	Positive Uniaxial
Type I	BPM	$n_e(2\omega,\theta) = n_o(\omega)$	$n_o(2\omega) = n_e(\omega,\theta)$
	GV matching	$n_e^{(g)}(2\omega,\theta) = n_o^{(g)}(\omega)$	$n_o^{(g)}(2\omega) = n_e^{(g)}(\omega,\theta)$
Type II	BPM	$2n_e(2\omega,\theta) = n_o(\omega) + n_e(\omega,\theta)$	$2n_o(2\omega) = n_o(\omega) + n_e(\omega,\theta)$
	GV matching	$2n_e^{(g)}(2\omega,\theta) = n_o^{(g)}(\omega) + n_e^{(g)}(\omega,\theta)$	$2n_o^{(g)}(2\omega) = n_o^{(g)}(\omega) + n_e^{(g)}(\omega,\theta)$

The orthorhombic ternary chalcogenides considered in this study (i.e., LIS, LISe, LGS, and LGSe) are all negative biaxial crystals belonging to the point group of orthorhombic $mm2$ at room temperature (see Table 1). Their crystallographic axes (a, b, and c) are all perpendicular to each other and have a relationship with the optical axes, where $(y, x, z) = (a, b, c)$, within the spectral region considered in this study [39,43–45]. In this assignment, the order of RI magnitudes is given by $n_x < n_y < n_z$. Then, the collinear BPM conditions can be expressed as:

$$\Delta k = \left| k^{(l)}(2\omega) - 2k^{(h)}(\omega) \right| = 0 \text{ (for type I, biaxial) and} \quad (11)$$

$$\Delta k = \left| k^{(l)}(2\omega) - k^{(h)}(\omega) - k^{(l)}(\omega) \right| = 0 \text{ (for type II, biaxial),} \quad (12)$$

where each k represents the wave number of the interacting wave and is defined as $k^{(m)}(j\omega) = (j\omega/c)n^{(m)}$. The unit k-vector is defined in the spherical coordinate as, (sinθcosφ, sinθsinφ, and cosθ). Here, θ and φ are the polar and azimuthal angles, respectively. The RIs of the two eigen-polarization modes of light traveling inside a biaxial crystal are expressed as follows by solving the Fresnel equation of the wave normal [46]:

$$n^{(m)}(j\omega) = \sqrt{\frac{2}{-B_j \pm \sqrt{B_j^2 - 4C_j}}}. \quad (13)$$

The parameters in Equation (13) are defined as follows:

$$B_j = -(b_j + c_j)k_x^2 - (a_j + c_j)k_y^2 - (b_j + a_j)k_z^2, \quad (14)$$

$$C_j = b_j c_j k_x^2 + a_j c_j k_y^2 + b_j a_j k_z^2, \text{ and} \quad (15)$$

$$a_j = n_x(j\omega), \ b_j = n_y(j\omega), \ c_j = n_z(j\omega), \quad (16)$$

where each k_i represents the x, y, and z-axis components of the wave vector. j can be 1 or 2 and then $n^{(m)}(\omega)$ and $n^{(m)}(2\omega)$ denote the RIs of the F-wave and the SH wave with frequencies ω and 2ω, respectively. m in Equation (13) can be either l or h, representing low or high RI. l and h are obtained by taking the plus and minus from the $\pm$ sign of the denominator in Equation (13), respectively. The GV matching conditions $\Delta n_g = 0$ obtained from Equation (6) are simplified as:

$$n_g^{(l)}(2\omega) = n_g^{(h)}(\omega) \text{ (for type I, biaxial) and} \quad (17)$$

$$2n_g^{(l)}(2\omega) = n_g^{(h)}(\omega) + n_g^{(l)}(\omega) \text{ (for type II, biaxial).} \quad (18)$$

Each subscript g in Equations (17) and (18) indicates the group index. Now, the broadband SHG in a biaxial crystal is defined as Equations (11) and (17) (for Type I) or Equations (12) and (18) (for Type II), being satisfied simultaneously. The BPM and GV matching conditions for broadband SHG in biaxial crystals are listed in Table 3. Each of these equations in Table 3 are a function of three variables: F-wavelength (λ_F), θ, and φ. Therefore, by solving the system of equations for BPM and GV matching while changing λ_F, we can get a set of solutions for θ and φ, i.e., the direction of the F-wave vector for broadband SHG at the given λ_F. Thus, the resonant λ_F can be continuously tuned by sweeping the F-wave vector along the direction characterized by the solution set of θ and φ. In other words, for the orthorhombic ternary chalcogenide considered in this study, the spectral position of the SH wave can be selectively determined or tuned within the range of the solution sets. In contrast, for chalcopyrite crystals exhibiting uniaxial birefringence, SH resonance can only be achieved at a single wavelength, as discussed earlier.

Table 3. The BPM and GV matching conditions for broadband SHG in biaxial crystals.

Type	Condition	Negative Biaxial
Type I	BPM	$n^{(l)}(2\omega) = n^{(h)}(\omega)$
	GV matching	$n_g^{(l)}(2\omega) = n_g^{(h)}(\omega)$
Type II	BPM	$2n^{(l)}(2\omega) = n^{(h)}(\omega) + n^{(l)}(\omega)$
	GV matching	$2n_g^{(l)}(2\omega) = n_g^{(h)}(\omega) + n_g^{(l)}(\omega)$

2.2. Effective Nonlinearities

The effective nonlinearity of a crystal depends on its point group and the NLO interaction type [47,48]. The point groups of chalcopyrite and defect chalcopyrite crystals considered in this study are $\bar{4}2m$, $\bar{4}$, $3m$, $\bar{6}2m$, and $6mm$ (see Table 1). The corresponding analytical expressions of the effective NLO coefficients (d_{eff}) are listed in column 5 of Table 4. In each d_{eff} expression, θ and φ represent the polar and azimuthal angles, respectively. The spatial walk-off angle (ρ) between the wave vector and the Poynting vector within a uniaxial crystal is derived as:

$$\tan\rho = \frac{\sin 2\theta}{2}\left(\frac{1}{n_e^2(\omega)} - \frac{1}{n_o^2(\omega)}\right)n_e^2(\omega,\theta), \quad (19)$$

where $n_e(\omega,\theta)$ is given in Equation (5). To derive this expression, we used the definition in [49]. Note that in each d_{eff} expression (column 5 of Table 4), θ is corrected as much as ρ (i.e., $\theta \to \theta+\rho$). As the interacting beams do not usually propagate along the crystallographic axis due to birefringence, there is a spatial walk-off between the beams even in the case of collinear BPM. Thus, we need to correct the angle from θ to $\theta+\rho$ to obtain d_{eff} accordingly. Considering the azimuthal angle has no effect on both BPM and GV matching conditions as shown in Table 2, φ can be chosen as an arbitrary value to maximize d_{eff}. As can be appreciated from Equation (19), ρ is also a function of λ_F and θ. Therefore, for uniaxial crystals, d_{eff} can be obtained by substituting the solution sets of λ_F and θ that satisfy broadband SHG conditions (i.e., Table 2). The NLO efficiency is proportional to the square of d_{eff} for a given direction of beam propagation [42]. The d_{il} components on the crystallographic axes were chosen as the measurements closest to the spectral region to be considered in this study and their references are also given in column 3 of Table 4.

For biaxial birefringent crystals such as orthorhombic ternary chalcogenides, the effective NLO coefficients for a given direction of the F-wave can generally be expressed as a linear combination of d_{il} components as follows:

$$d_{eff}^{(t)} = \xi_1^{(t)}d_{15} + \xi_2^{(t)}d_{24} + \xi_3^{(t)}d_{31} + \xi_4^{(t)}d_{32} + \xi_5^{(t)}d_{33}, \quad (20)$$

where the superscript t can be either I or II and represents Type I and Type II NLO interactions, respectively. The d_{il} components on the crystallographic axes were chosen as the measurements closest to the spectral region to be considered in this study and the values are given in Table 5 with references. Each ξ coefficient in Equation (20) is determined by the relationship between the optical axes and the crystallographic axes of a biaxial crystal [48]. For $mm2$ crystals (e.g., LIS, LISe, LGS, and LGSe) showing the relationships of $(y, x, z) = (a, b, c)$, the ξ-coefficients are given as follows:

Table 4. Effective NLO coefficients (d_{eff}) of chalcopyrite crystals showing uniaxial birefringence.

Crystals	Point Group	d_{il} Component (pm/V)	BPM Type	d_{eff} Expression [1]
AGS	$\bar{4}2m$	$d_{36} = 13.7$ (@ $\lambda_F = 2.53$ μm [50])	Type I Type II	$-d_{36}\sin(\theta+\rho)\sin 2\varphi$ $d_{36}\sin[2(\theta+\rho)]\cos 2\varphi$
AGSe	$\bar{4}2m$	$d_{36} = 33$ (@ $\lambda_F = 2.3$ μm [51])	Type I Type II	$-d_{36}\sin(\theta+\rho)\sin 2\varphi$ $d_{36}\sin[2(\theta+\rho)]\cos 2\varphi$
CSP	$\bar{4}2m$	$d_{36} = 84.5$ (@ $\lambda_F = 4.56$ μm [52])	Type I Type II	$-d_{36}\sin(\theta+\rho)\sin 2\varphi$ $d_{36}\sin[2(\theta+\rho)]\cos 2\varphi$
HGS	$\bar{4}$	$d_{36} = 31.5, d_{31} = 10.5$ (@ $\lambda_F = 1.064$ μm [53])	Type I Type II	$\sin(\theta+\rho)(d_{36}\sin 2\varphi + d_{31}\cos 2\varphi)$ $\sin[2(\theta+\rho)](d_{36}\cos 2\varphi - d_{31}\sin 2\varphi)$
TASe	$3m$	$d_{22} = 32, d_{31} = 20$ (@ $\lambda_F = 10.6$ μm [54])	Type I Type II	$d_{15}\sin(\theta+\rho) - d_{22}\cos(\theta+\rho)\sin 3\varphi$ $d_{22}\cos^2(\theta+\rho)\cos 3\varphi$
GaSe	$\bar{6}2m$	$d_{22} = 54$ (@ $\lambda_F = 10.6$ μm [51])	Type I Type II	$-d_{22}\cos(\theta+\rho)\sin 3\varphi$ $d_{22}\cos^2(\theta+\rho)\cos 3\varphi$
CGA	$\bar{4}2m$	$d_{36} = 186$ (@ $\lambda_F = 5.2955$ μm [55])	Type I Type II	$d_{36}\sin[2(\theta+\rho)]\cos 2\varphi$ $-d_{36}\sin(\theta+\rho)\sin 2\varphi$
ZGP	$\bar{4}2m$	$d_{36} = 70$ (@ $\lambda_F = 5.2955$ μm [56])	Type I Type II	$d_{36}\sin[2(\theta+\rho)]\cos 2\varphi$ $-d_{36}\sin(\theta+\rho)\sin 2\varphi$
CdSe	$6mm$	$d_{31} = 18$ (@ $\lambda_F = 10.6$ μm [51])	Type I Type II	0 $d_{31}\sin(\theta+\rho)$

[1] The walk-off angle ρ is given as a function of the polar angle θ at resonance (see Equation (19)).

$$\begin{aligned}
\zeta_1^{(I)} &= 2AH(BCH + EG)(BCE - GH), \\
\zeta_2^{(I)} &= 2AH(BEG + CH)(BGH - CE), \\
\zeta_3^{(I)} &= AE(BCH + EG)^2, \\
\zeta_4^{(I)} &= AE(BGH - CE)^2, \\
\zeta_5^{(I)} &= A^3H^2E,
\end{aligned} \quad (21)$$

$$\begin{aligned}
\zeta_1^{(II)} &= -AE(BCH + GE)(BCE - GH) - AH(BCE - GH)^2, \\
\zeta_2^{(II)} &= -AE(BGE + CH)(BGH - CE) - AH(BGE + CH)^2, \\
\zeta_3^{(II)} &= -AE(BCH + GE)(BCE - GH), \\
\zeta_4^{(II)} &= -AE(BGE + CH)(BGH - CE), \\
\zeta_5^{(II)} &= -A^3E^2H,
\end{aligned} \quad (22)$$

where the angle-dependent parameters of A, B, C, G, E, and H represent $\sin\theta$, $\cos\theta$, $\sin\varphi$, $\cos\varphi$, $\sin\delta$, and $\cos\delta$, respectively. The angle δ introduced for convenience only is defined as:

$$\tan\delta \equiv \frac{2BGC}{A^2\cot^2 V_z - B^2G^2 + C^2}, \quad (23)$$

where V_z represents the angle between the z-axis and the optic axis of biaxial birefringence. For biaxial crystals with a relationship of $n_x < n_y < n_z$, such as all four chalcogenides considered in this study, V_z is given by:

$$\sin V_z = \frac{n_z(n_y^2 - n_x^2)^{1/2}}{n_y(n_z^2 - n_x^2)^{1/2}}. \quad (24)$$

As V_z is a function of RI, d_{eff} is given as a function of the three variables λ_F, θ, and φ. Thus, d_{eff} can be obtained by substituting the set of solutions for λ_F, θ, and φ, satisfying the broadband SHG conditions (i.e., Table 3). The SHG efficiency is proportional to the square of d_{eff} for a given direction of beam propagation [42].

Table 5. The d_{il} components of orthorhombic ternary chalcogenides considered in this study.

Crystals	Point Group	d_{il} Components (pm/V) [1]
LIS	mm2	$d_{31} = 7.2 \pm 0.4$, $d_{32} = 5.7 \pm 0.6$, $d_{33} = -16 \pm 4$ (@ $\lambda_F = 2.53$ μm [57])
LISe		$d_{31} = 11.78 \pm 5\%$, $d_{32} = 8.17 \pm 10\%$, $d_{33} = -16 \pm 25\%$ (@ $\lambda_F = 2.53$ μm [40])
LGS		$d_{31} = 5.8$, $d_{32} = 5.1$, $d_{33} = -10.7$ (@ $\lambda_F = 2.53$ μm [58])
LGSe		$d_{31} = 9.9$, $d_{32} = 7.7$, $d_{33} = -18.2$ (@ $\lambda_F = 2.53$ μm [58])

[1] $d_{31} = d_{15}$ and $d_{32} = d_{24}$ under Kleinman symmetry.

2.3. Spatial Walk-Off

If the wave vector of the input light is not parallel to one of the crystallographic axes, a spatial walk-off occurs between the wave vector and the Poynting vector inside the crystal, as described in Equation (19). For uniaxial crystals, ρ corresponds to the angle between the o-wave and e-wave [42]. Then, the largest walk-off angle (w) between the SH-beam and one of the two F-beams can be defined as the maximum angle between the interacting o and e-waves. As can be appreciated from the BPM conditions in Table 2, w is equal to ρ_ω (for positive uniaxial crystals) or $\rho_{2\omega}$ (for negative uniaxial crystals), where ρ_ω and $\rho_{2\omega}$ represent the walk-off angles of the e-waves with frequencies ω and 2ω, respectively. We note that for Type II, the relationship of $\rho_{2\omega} > \rho_\omega$ is valid in the uniaxial crystals considered in this study [59]. As expected from Equation (19), w is also given as a two-variable function for λ_F and θ, and thus can be obtained by substituting a set of solutions for λ_F and θ that satisfy the broadband SHG conditions in Table 2.

For biaxial crystals with the point symmetry mm2, the walk-off angle is expressed as:

$$\tan \rho_j^{(m)} \equiv \left\{ n^{(m)}(j\omega) \right\}^2 \left[\left(\frac{k_x}{\left\{ n^{(m)}(j\omega) \right\}^{-2} - a_j} \right)^2 + \left(\frac{k_y}{\left\{ n^{(m)}(j\omega) \right\}^{-2} - b_j} \right)^2 + \left(\frac{k_z}{\left\{ n^{(m)}(j\omega) \right\}^{-2} - c_j} \right)^2 \right]^{-1/2} \quad (25)$$

where all parameters of Equation (25) are defined in Equations (13)–(16). For Type I, w is simply determined by the angle between the low-RI SH-beam and the high-RI F-beam, as can be appreciated from Table 3. However, for Type II, the low-RI F-beam is always placed between the high-RI F-beam and the low-RI SH-beam [60,61]. In this case, w is defined as the largest angle formed by the high-RI F-beam and the low-RI SH-beam:

$$\cos w = \cos \rho_1^{(h)} \cos \rho_2^{(l)}. \quad (26)$$

Now, w is also given as a function of the three variables λ_F, θ, and φ, which can be obtained by substituting the solution sets of λ_F, θ, and φ, satisfying the broadband SHG into Equation (26). The maximum deviation between the interacting beams after passing through an NLO crystal of length L can be expressed as:

$$\Delta = L \tan w. \quad (27)$$

3. Simulations and Discussion

3.1. Broadband SHG in Uniaxial Chalcopyrite and Defect Chalcopyrite Crystals

Figure 2 shows the BPM properties of negative uniaxial crystals among chalcopyrite semiconductors and defect chalcopyrite, namely AGS, AGSe, CSP, HGS, TASe, and GaSe (see Table 1). The red and blue surfaces represent $n_e(\lambda_F/2, \theta)$ and $n_o(\lambda_F)$ (for Type I) or $2n_e(\lambda_F/2, \theta)$ and $n_o(\lambda_F) + n_e(\lambda_F, \theta)$ (for Type II), as expected from Table 2. The base of the coordinates is the plane formed by the angles θ and λ_F, indicating the direction of the F-wave vector and the wavelength, respectively. Each intersection of the two surfaces in Figure 2 indicates that the BPM is possible over a specific range of λ_F. The results show that both Type I and Type II NLO interactions are possible in all considered cases of AGS,

AGSe, CSP, HGS, TASe, and GaSe. Figure 3 shows the BPM properties of chalcopyrite semiconductors and defect chalcopyrite exhibiting positive uniaxial birefringence (i.e., CGA, ZGP, and CdSe, as listed in Table 1). In this case, the red and blue surfaces represent $n_o(\lambda_F/2)$ and $n_e(\lambda_F,\theta)$ (for Type I) or $2n_o(\lambda_F/2)$ and $n_o(\lambda_F) + n_e(\lambda_F,\theta)$ (for Type II). As shown in Figure 3d,f, the two surfaces do not intersect and thus Type II interactions are not possible for ZGP and CdSe. Therefore, for chalcopyrite crystals with positive uniaxial birefringence (see Table 1), two types of BPM are possible in CGA, whereas only Type I interactions are possible in ZGP and CdSe. However, as the effective nonlinearity of CdSe for Type I is zero (i.e., d_{eff} = 0 in Table 4), consequently, SHG is not possible for both Types I and II in CdSe.

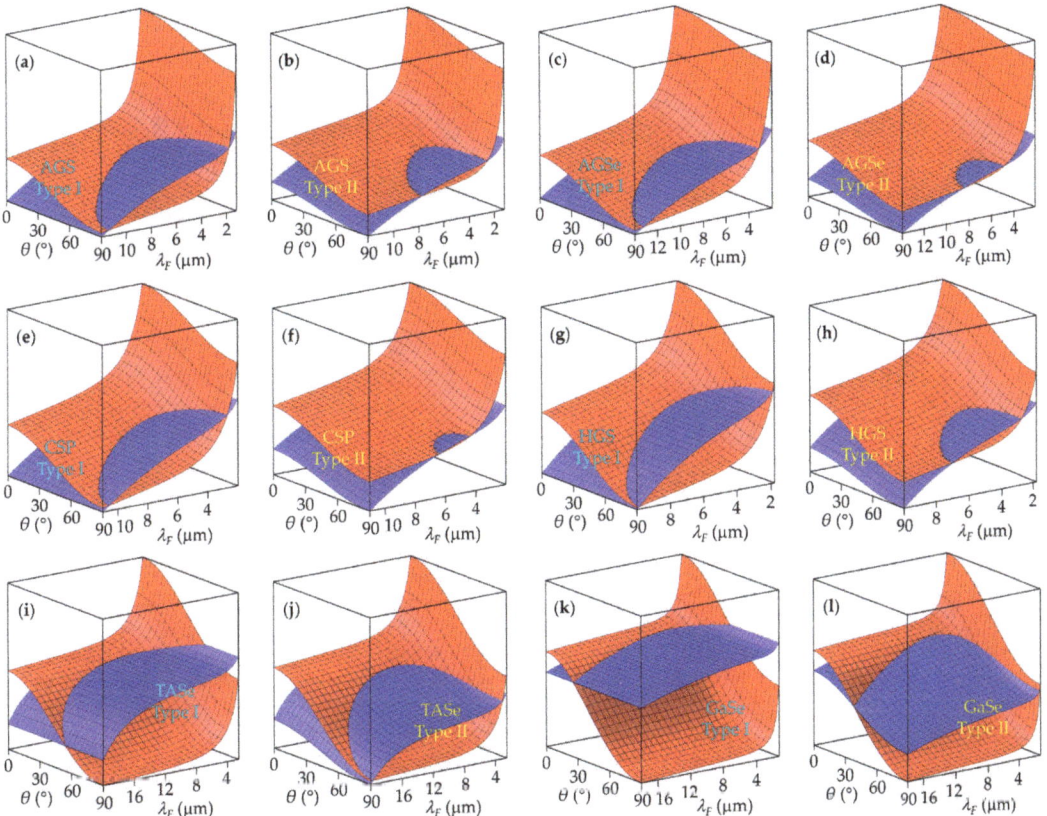

Figure 2. BPM properties of negative uniaxial crystals among chalcopyrite semiconductors and defect chalcopyrite exhibiting negative uniaxial birefringence: (**a**) Type I and (**b**) Type II in AGS; (**c**) Type I and (**d**) Type II in AGSe; (**e**) Type I and (**f**) Type II in CSP; (**g**) Type I and (**h**) Type II in HGS; (**i**) Type I and (**j**) Type II in TASe; and (**k**) Type I and (**l**) Type II in GaSe. The red and blue surfaces represent $n_e(\lambda_F/2,\theta)$ and $n_o(\lambda_F)$ (for Type I) or $2n_e(\lambda_F/2,\theta)$ and $n_o(\lambda_F) + n_e(\lambda_F,\theta)$ (for Type II).

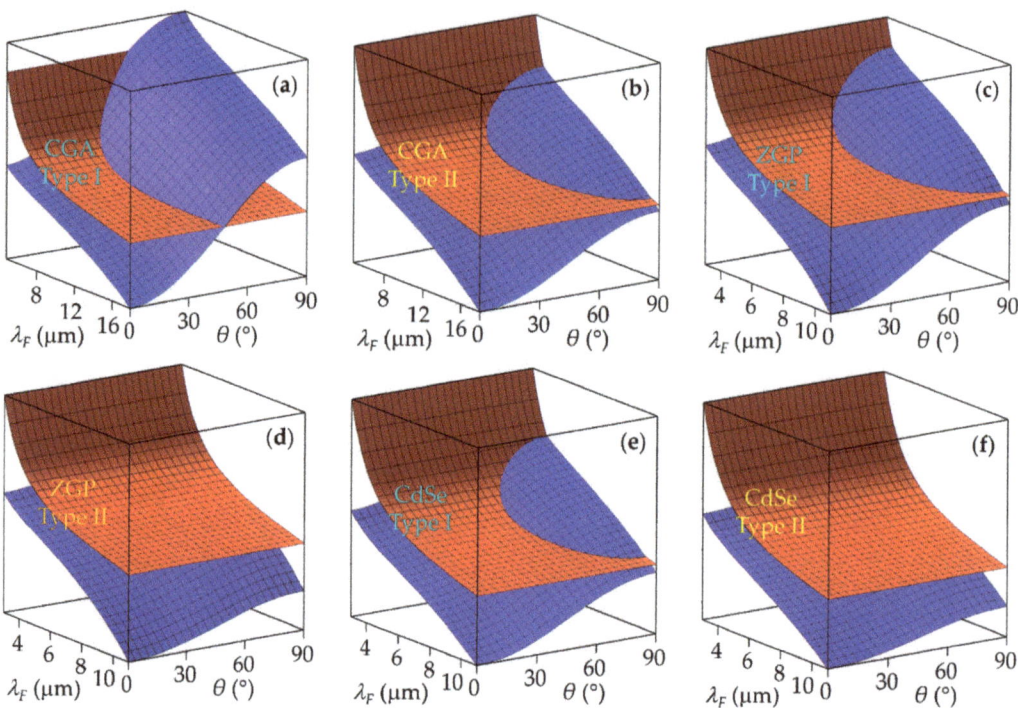

Figure 3. BPM properties of chalcopyrite semiconductors and defect chalcopyrite exhibiting positive uniaxial birefringence: (**a**) Type I and (**b**) Type II in CGA; (**c**) Type I and (**d**) Type II in ZGP; and (**e**) Type I and (**f**) Type II in CdSe. The red and blue surfaces represent $n_o(\lambda_F/2)$ and $n_e(\lambda_F,\theta)$ (for Type I) or $2n_o(\lambda_F/2)$ and $n_o(\lambda_F) + n_e(\lambda_F,\theta)$ (for Type II).

Figure 4 shows the numerical simulation results of the BPM and GV matching for the broadband SHG: Figure 4a–h correspond to AGS, AGSe, CSP, HGS, TASe, GaSe, CGA, and ZGP, respectively. The solid red and blue lines in each graph represent the BPM curves for Type I and Type II, respectively, corresponding to the intersection lines in Figures 2 and 3. The dashed magenta and cyan lines in each graph correspond to the GV matching curves for Type I and Type II, respectively, which are plotted using Equations (7)–(10). The intersection of the red and magenta (or blue and cyan) curves in Figure 4 indicates the specific direction of the F-wave vector (i.e., θ_{BPM}) and the λ_F value corresponding to the SH resonance for Type I (or Type II). As discussed earlier, for ZGP, the BPM and GV matching curves intersect only for Type I (see Figure 4g). The F-wave resonances for the broadband SHG, the corresponding SH wavelengths, and the directions of the F-wave vector for the BPM obtained at the intersections in Figure 4 are summarized in Table 6. The Sellmeier equations of each crystal used in the calculations are listed together in Table 6. The F-wave resonances are interspersed in the spectral range of 4.37–9.47 µm, which corresponds to the spectral range of various mid-IR lasers such as high-power quantum cascade lasers (QCLs) based on buried-ridge or strain-balanced waveguides (WGs); distributed feedback (DFB) lasers based on plasmon-enhanced WGs or corrugated surface gratings; external cavity lasers in various bound-to-continuum designs; solid state lasers based on chalcogenide crystals doped with Fe^{2+}; and gas lasers [62–75].

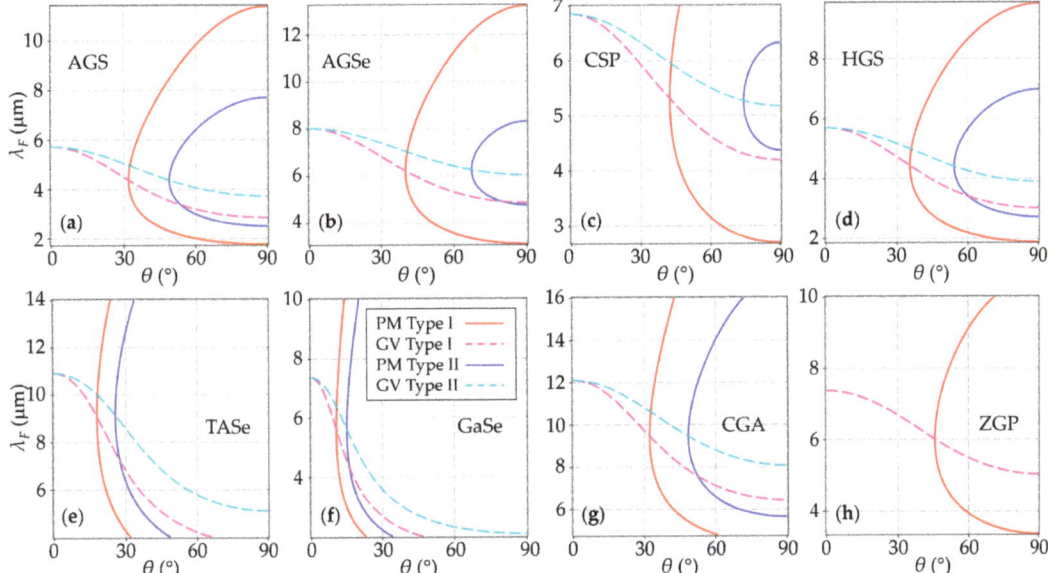

Figure 4. BPM and GV matching curves of (**a**) AGS, (**b**) AGSe, (**c**) CSP, (**d**) HGS, (**e**) TASe, (**f**) GaSe, (**g**) CGA, and (**h**) ZGP. The intersection of the red and magenta (or blue and cyan) curves indicates the direction of the F-wave vector (i.e., θ) and the λ_F value corresponding to the SH resonance for Type I (or Type II).

Table 6. Broadband SHG conditions in mid-IR uniaxial crystals: the resonant F-wavelengths (λ_F) and the corresponding SH wavelengths, (λ_{SH}); the BPM direction (θ_{BPM}); the effective NLO coefficients (d_{eff}); the maximum walk-off angles between the interacting waves (w); and the acceptable F-bandwidths for the broadband SHG ($\Delta\lambda_F$).

Crystals	BPM Type	λ_F (nm)	λ_{SH} (nm)	θ_{BPM} (°)	$\lvert d_{eff} \rvert$ (pm/V) [1]	w (°)	$\Delta\lambda_F$ (nm)
AGS [76]	Type I	4386.04	2193.02	31.92	7.5	1.16	488.69
	Type II	4373.83	2186.92	48.87	13.5	1.27	484.71
AGSe [77]	Type I	6224.24	3112.12	39.93	21.5	0.70	870.19
	Type II	6271.35	3135.67	67.21	23.2	0.50	868.39
CSP [52]	Type I	5315.10	2657.55	42.36	57.5	0.53	697.04
	Type II	5270.76	2635.38	74.35	43.2	0.27	669.20
HGS [78]	Type I	4438.55	2219.28	35.16	27.2	0.94	506.49
	Type II	4412.72	2206.36	54.18	31.2	0.93	506.54
TASe [35]	Type I	9076.20	4538.10	17.95	18.6	2.12	1122.97
	Type II	9140.37	4570.18	25.64	24.7	2.79	1145.78
GaSe [79]	Type I	5625.28	2812.64	10.57	52.5	2.99	760.91
	Type II	5620.10	2810.05	15.08	48.2	4.12	760.70
CGA [55]	Type I	9474.36	4737.18	32.13	163.6	−1.33	1164.03
	Type II	9474.36	4737.18	48.44	135.9	−1.49	1164.24
ZGP [80]	Type I	6006.92	3003.46	45.75	70.0	−0.70	644.54
	Type II	-	-	-	-	-	-
CdSe [81]	Type I	6380.33	3190.17	53.04	0	−0.44	-
	Type II	-	-	-	-	-	-

[1] The tensor components (d_{ij}) used for the calculations are listed in Table 4.

Considering the NLO efficiency is proportional to the square of d_{eff}, it is critical to estimate the effective nonlinearity for the given direction of the F-wave vector that satisfies the broadband SHG. Figure 5 shows the polar-plots of the d_{eff} values as a function of θ, which were numerically calculated using the analytical equations for chalcopyrite uniaxial crystals in Table 4. The walk-off effects in Equation (19) were considered in the calculation results as discussed in Table 4. As shown in Figure 5, each polar-plot of d_{eff} exhibits two-fold or four-fold rotational symmetry with respect to θ, which is determined by the point symmetry of the crystal. The d_{eff} values calculated for the directions (θ_{BPM}) of the F-wave satisfying the broadband SHG conditions are listed in column 6 of Table 6. For all considered cases except CdSe, the d_{eff} values are very large, from 7.5 pm/V to 163.6 pm/V, which are much larger than the maximum d_{eff} for typical 5-mol% MgO-doped periodically poled LiNbO$_3$ (MgO:PPLN) using the first-order QPM: $(2/\pi)d_{31}$ = ~2.80 pm/V for Type I and II [59]. The walk-off angle w, calculated using the solution set of λ_F and θ for the broadband SHG, are listed in column 7 of Table 6. The calculated values of w are either positive (for negative uniaxial crystals) or negative (for positive uniaxial crystals) because the deviation of the e-ray with respect to the o-ray is opposite depending on the type of uniaxial crystal. The estimated walk-offs can be sufficiently overcome by a large beam window in thick crystals. The calculated values of d_{eff} and w are also summarized in Table 6 along with other conditions for the broadband SHG.

Now, we will discuss the acceptable bandwidth of the F-wave for the broadband SHG. Figure 6 shows the spectra of F-waves that are acceptable for the broadband SHG in uniaxial mid-IR crystals, in which the solid red and dashed blue lines indicate Type I and II, respectively. The spectra of F-waves were plotted using the well-known spectral equation in the coupled mode theory, in which the NLO interaction length used in the calculation is 10 mm [42]. In the proposed simultaneous BPM–GV matching scheme, the crystal length is not limited by the temporal walk-off between the interacting wave due to the zero GVM. A crystal length of 10 mm was chosen as a reference value for comparison. In the F-wave spectra of GaSe and CGA shown in Figure 6f,g, respectively, the curves for Types I and II almost overlap because the spectral positions of the resonances and their bandwidths in the two types are almost the same. In the case of ZGP, as described earlier, Type II does not exist, thus only the red curve corresponding to Type I is plotted in Figure 6h. The center wavelength (λ_F) of each graph in Figure 6 and the calculated bandwidths ($\Delta\lambda_F$) are summarized Table 6. The calculated bandwidths span from 488.69 nm to 1164.24 nm in full-width-half-maximum (FWHM) as listed in Table 6. Considering such broad spectral bandwidth can cover the full spectral width of sub-picosecond pulses, as shown in Figure 6, the SHG process can be used for continuum generation, as expected in [12,30]. For example, assuming transform-limited Gaussian pulses, the calculated input bandwidths correspond to the temporal widths ranging from 56.4 fs to 113.2 fs. In particular, parametric up-conversion using short pulse trains can potentially be applied to optical imaging and microscopy in the mid-IR region, in information transmission, and nonlinear optical signal processing. [27–29].

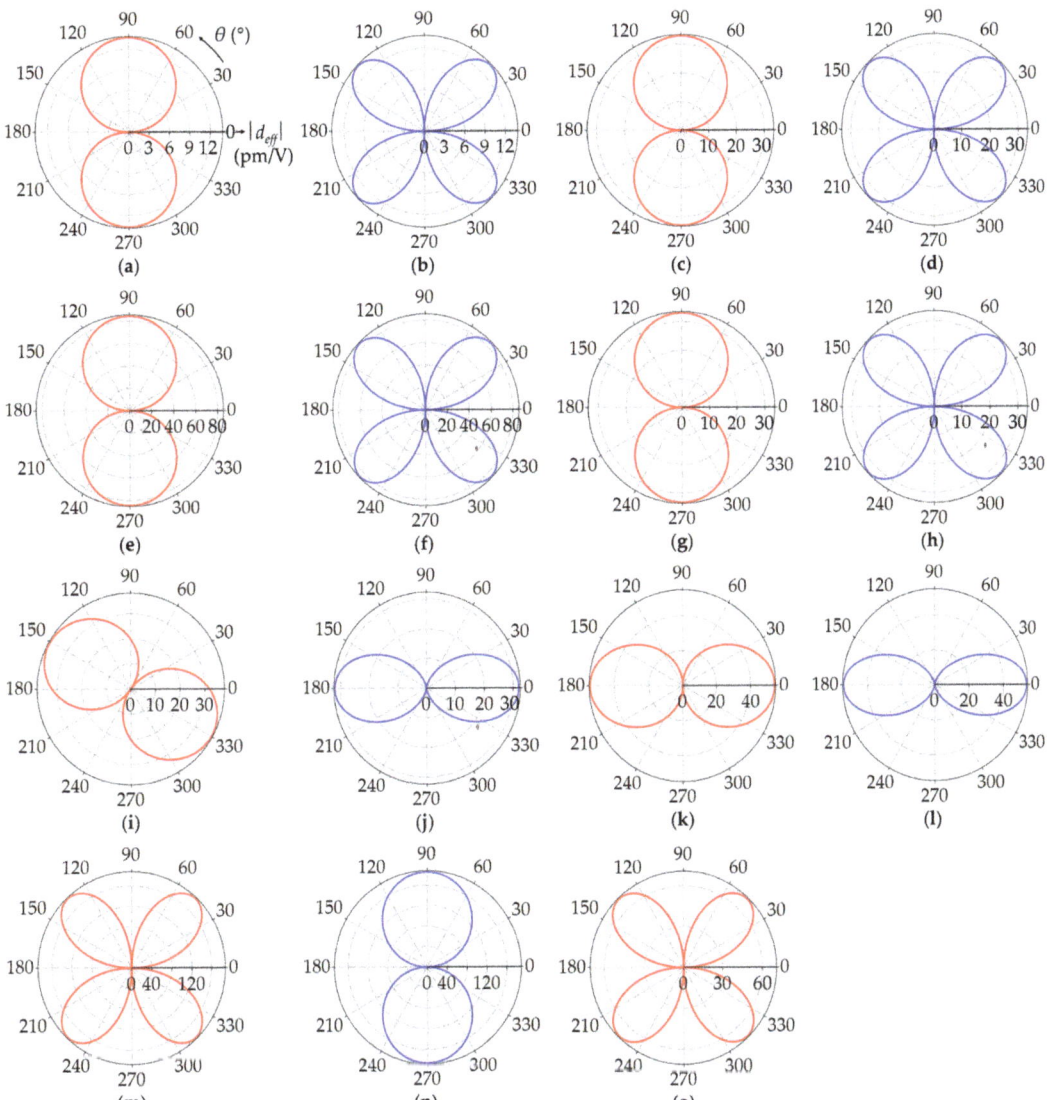

Figure 5. The polar-plots of the d_{eff} values shown as a function of θ, which were numerically calculated using the analytical equations for chalcopyrite uniaxial crystals in Table 4. (**a**) Type I and (**b**) Type II in AGS; (**c**) Type I and (**d**) Type II in AGSe; (**e**) Type I and (**f**) Type II in CSP; (**g**) Type I and (**h**) Type II in HGS; (**i**) Type I and (**j**) Type II in TASe; (**k**) Type I and (**l**) Type II in GaSe; (**m**) Type I and (**n**) Type II in CGA; and (**o**) Type I in ZGP.

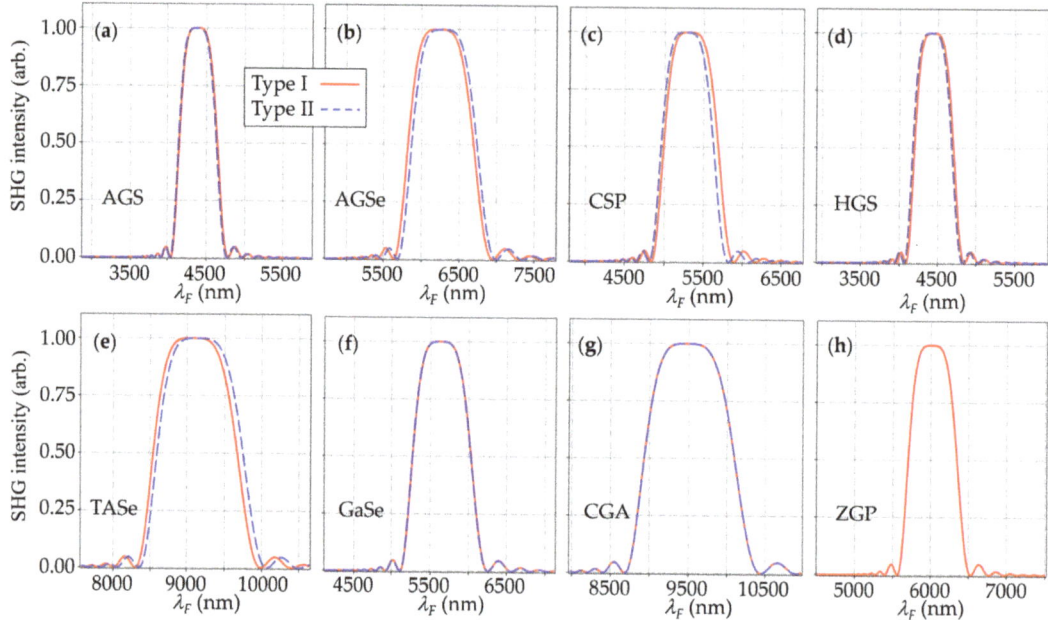

Figure 6. The spectra of F-waves that are acceptable for broadband SHG in uniaxial mid-IR crystals: (**a**) AGS, (**b**) AGSe, (**c**) CSP, (**d**) HGS, (**e**) TASe, (**f**) GaSe, (**g**) CGA, and (**h**) ZGP. Solid red line: Type I; dashed blue line: Type II.

3.2. Broadband SHG in Biaxial Orthorhombic Ternary Chalcogenides

Figure 7 shows the BPM and GV matching properties of biaxial orthorhombic ternary chalcogenides, namely LIS, LISe, LGS, and LGSe in Table 1. The vertical axis of λ_F covers the mid-IR spectral range and the base of the coordinates is the plane formed by the angles θ and φ representing the direction of the F-wave vector. The red and magenta (or blue and cyan) surfaces in each graph indicate the BPM and GV matching surfaces for Type I (or Type II, respectively), which were calculated using the equations in Table 3. In each graph in Figure 7, the intersection line of the two surfaces spans a specific range of λ_F, θ, and φ, where the BPM and GV matching is satisfied simultaneously (i.e., the broadband SHG condition). This means that the spectral position of the F-wave resonance can be selectively determined or tuned within that range of λ_F, θ, and φ, satisfying the broadband SHG conditions. In contrast, for uniaxial mid-IR crystals such as chalcopyrite semiconductors, the F-wave resonance can only be achieved at a single wavelength, as described in Section 3.1.

Figure 8 plots the directions of the F-wave vector (i.e., θ and φ) as functions of the resonant λ_F, corresponding to the intersection lines in Figure 7. As can be seen from Figure 8, for both types, the F-wave vector satisfying the broadband SHG conditions deviates more from the optical x and z-axis at shorter wavelengths. The F-wave resonance (λ_F) ranges, corresponding SH wavelength (λ_{SH}) ranges, and F-wave vector directions (θ and φ) for the broadband SHG are listed in Table 7 by the BPM type of each crystal. The Sellmeier equations for the mid-IR biaxial crystal used in the calculations are listed together in Table 7. The F-wave resonances span over the spectral range of 3.5–5.1 µm, which corresponds to the spectral range of mid-IR lasers such as high-power QCLs, DFB lasers, optical parametric oscillator lasers, solid state crystalline lasers, and gas lasers [65,66,70–74,82–85].

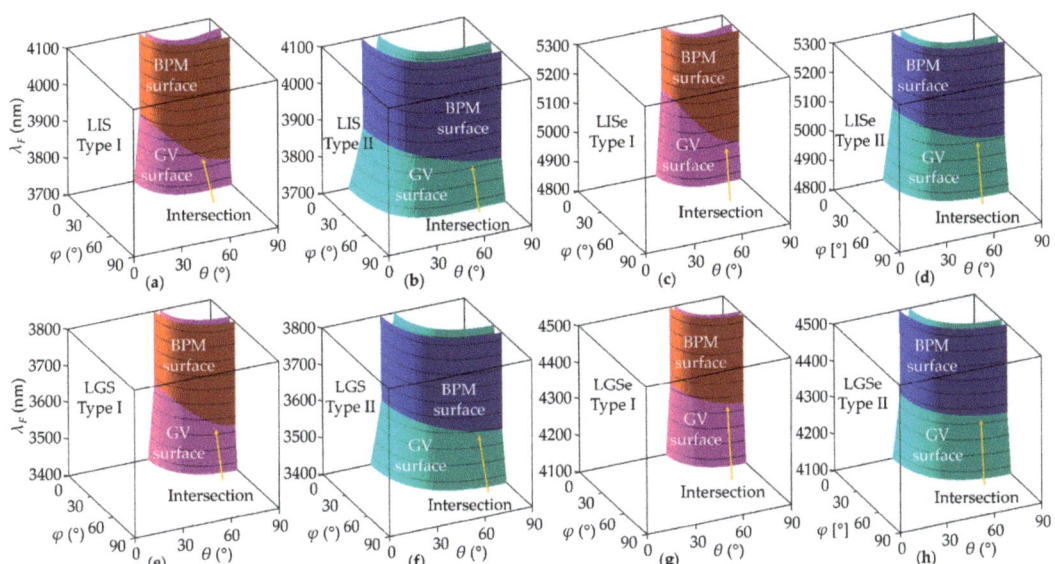

Figure 7. Numerical simulation results of BPM and GV matching in biaxial orthorhombic ternary chalcogenides: (**a**) Type I and (**b**) Type II in LIS; (**c**) Type I and (**d**) Type II in LISe; (**e**) Type I and (**f**) Type II in LGS; and (**g**) Type I and (**h**) Type II in LGSe.

Table 7. Broadband SHG conditions of mid-IR biaxial crystals: the ranges of λ_F and λ_{SH}, and the corresponding range of the F-wave vector direction (θ and φ).

Crystals	BPM Type	λ_F Range (nm)	λ_{SH} Range (nm)	φ Range (°)	θ Range (°)
LIS [39]	Type I	3769.29—3879.91	1884.64—1939.95	26.10—0.00	90.00—45.57
	Type II	3809.60—3856.09	1904.80—1928.04	49.39—0.00	90.00—25.75
LISe [43]	Type I	4926.06—5095.49	2463.03—2547.75	22.31—0.00	90.00—51.91
	Type II	4981.24—5039.20	2490.62—2519.60	40.71—0.00	90.00—37.53
LGS [44]	Type I	3517.15—3601.95	1758.58—1800.98	31.69—0.00	90.00—54.48
	Type II	3535.79—3567.88	1767.90—1783.94	49.69—0.00	90.00—37.22
LGSe [45]	Type I	4259.98—4288.26	2129.99—2144.13	27.21—0.00	90.00—56.60
	Type II	4258.53—4269.03	2129.26—2134.52	43.32—0.00	90.00—42.01

The magnitudes of the effective NLO coefficients (d_{eff}) of biaxial crystals are determined by the set of solutions for λ_F, θ, and φ, satisfying the broadband SHG conditions, as described in the paragraphs with Equations (20)–(24). We note, again, that it is important to estimate the d_{eff} values for the given conditions because the efficiency of SHG is proportional to the square of d_{eff}. The d_{eff} values were calculated numerically using Equations (20)–(24) and the results are plotted in Figure 9 as functions of the resonant λ_F, corresponding to the horizontal axis in each graph in Figure 8. This region of λ_F corresponds to the spectral region of the F-wave resonance that satisfies the broadband SHG conditions (i.e., the intersection lines shown in Figure 7). The solid red and blue curves in Figure 9 represent the calculated d_{eff} values for Type I and II, respectively. For each interaction type, the change in the d_{eff} value with the increasing λ_F shows a similar trend for the four kinds of biaxial crystals (LIS, LISe, LGS, and LGSe). It is interesting to note that for Type II, the d_{eff} value is greatest when the direction of the F-wave vector is in the x-y plane (i.e., $\theta = 90°$), as shown Figure 9 and in column 6 of Table 7. In Figure 9, the λ_F values obtained at the peak points of each graph are summarized in Table 8, in addition to the corresponding λ_{SH}, the direction of the F-wave vector (θ and φ), and the d_{eff} values. As can be seen from Table 8, the d_{eff} values calculated for the orthorhombic ternary chalcogenides are still larger than that of MgO:PPLN (~2.8 pm/V), as discussed in Section 3.1, but overall are slightly smaller than those of the chalcopyrite crystals (see Table 6).

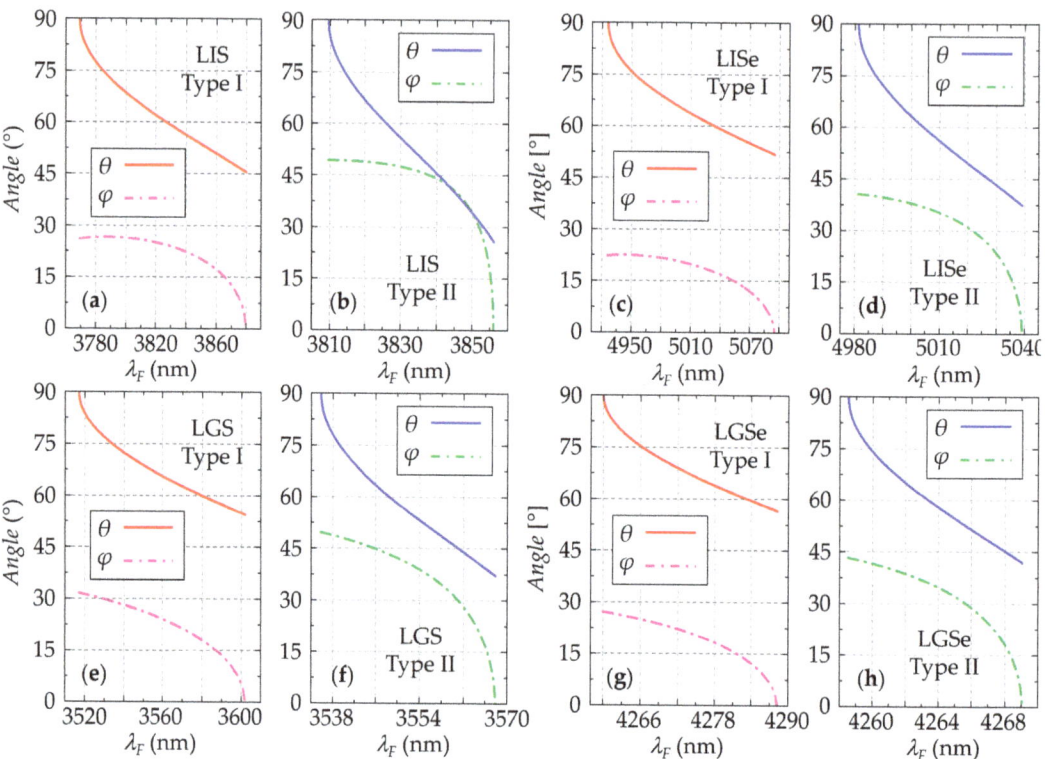

Figure 8. The directions (θ and φ) of F-wave vectors plotted as functions of λ_F satisfying the broadband SHG conditions. (**a**) Type I and (**b**) Type II in LIS; (**c**) Type I and (**d**) Type II in LISe; (**e**) Type I and (**f**) Type II in LGS; and (**g**) Type I and (**h**) Type II in LGSe.

The maximum walk-off angles (w) between the interacting waves calculated at the maximum d_{eff} value points (in Figure 9) using Equation (26) are listed in column 8 of Table 8. The calculated w values are less than 1.25° for all considered cases. The corresponding maximum beam deviation (Δ) calculated using Equations (27) is 22.8 µm/mm, which can be sufficiently overcome by using a larger-size F-beam in thick crystals. Considering these values are smaller than those of other widely used NLO crystals (e.g., w = 3.5° and Δ = 61.16 µm/mm for BBO), longer NLO interaction lengths within crystals can be used for higher SHG efficiency [42]. In particular, the efficient SHG with these small spatial-walk-offs is very desirable for terahertz generation from mid-infrared two-color laser filaments [86,87].

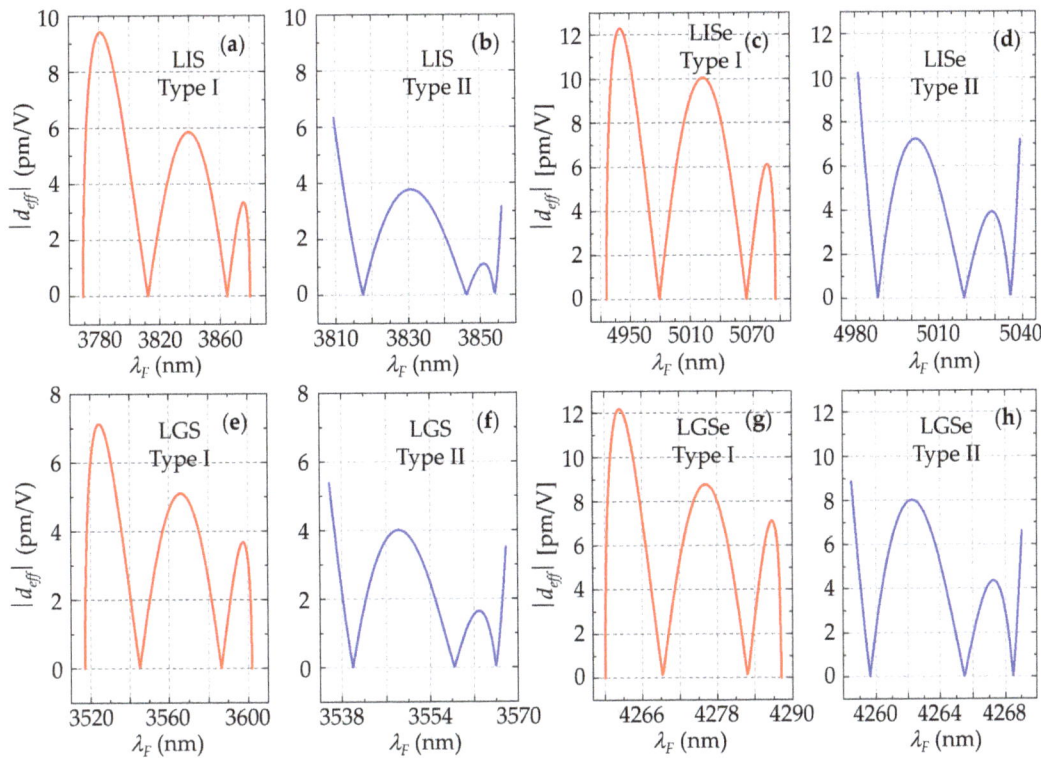

Figure 9. The effective NLO coefficient (d_{eff}) plotted as functions of λ_F satisfying the broadband SHG conditions. (**a**) Type I and (**b**) Type II in LIS; (**c**) Type I and (**d**) Type II in LISe; (**e**) Type I and (**f**) Type II in LGS; and (**g**) Type I and (**h**) Type II in LGSe.

Table 8. Broadband SHG conditions for the maximum d_{eff}: the resonant F-wavelength (λ_F), the direction of the F-wave vector (i.e., θ and φ), the effective NLO coefficient (d_{eff}), the walk-off angle (w), and the acceptable bandwidth of the F-wave ($\Delta\lambda_F$).

Crystals	BPM Type	λ_F (nm)	λ_{SH} (nm)	φ (°)	θ (°)	$\|d_{eff}\|$ (pm/V) [1]	w (°)	$\Delta\lambda_F$ (nm)
LIS	Type I	3781.02	1890.51	26.58	76.84	9.4	0.86	402.05
	Type II	3809.60	1904.00	19.39	90.00	6.3	0.94	410.30
LISe	Type I	4940.40	2470.20	22.55	78.97	12.3	0.82	629.79
	Type II	4981.24	2490.62	40.71	90.00	10.2	1.01	648.12
LGS	Type I	3524.58	1762.29	30.67	80.07	7.1	1.03	369.30
	Type II	3535.79	1767.90	49.69	90.00	5.4	1.11	375.76
LGSe	Type I	4262.46	2131.23	26.41	80.65	12.2	1.06	488.26
	Type II	4258.53	2129.26	43.32	90.00	8.9	1.25	491.01

[1] The tensor components (d_{il}) used for the calculations are listed in Table 5.

Finally, we describe the spectral bandwidths of F-waves allowed for broadband SHG in LIS, LISe, LGS, and LGSe. Figure 10 shows the F-wave spectra acceptable for the broadband SHG, in which the solid red and dashed blue curves indicate the spectra for Type I and II, respectively. The graphs were plotted using the well-known NLO coupled mode theory, in which the NLO interaction length of 10 mm was used in the calculations [42]. Each graph in Figure 10 is plotted for the case where d_{eff} has a maximum value in each crystal (i.e., the cases listed in Table 8). The spectral positions of F-wave resonances (λ_F) in each graph of Figure 10 and the calculated bandwidths ($\Delta\lambda_F$) are summarized in Table 8. The calculated bandwidths span from 369.30 nm to 648.12 nm in FWHM as listed in Table 8. In contrast to uniaxial crystals, the spectral position of the resonance in biaxial crystals can be tuned by sweeping the F-wave vector within the ranges of θ and φ as in Figure 8, while maintaining the broad bandwidth. As in the cases of uniaxial chalcopyrite crystals, such broad spectral bandwidth can cover the full spectral width of sub-picosecond pulses, allowing the SHG process to be used for continuum generation [12,30]. In particular, parametric up-conversion using short pulse trains can potentially be applied to optical imaging and microscopy in the mid-IR region, in information transmission using optical pulse signals, and in nonlinear optical signal processing [27–29].

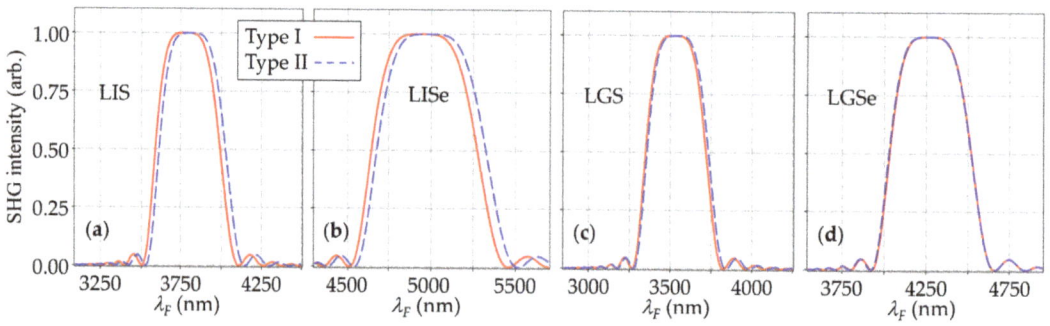

Figure 10. The spectra of F-waves that are acceptable for broadband SHG in biaxial mid-IR crystals: (**a**) LIS, (**b**) LISe, (**c**) LGS, and (**d**) LGSe. Solid red line: Type I; dashed blue line: Type II.

We note that the spectra plotted in Figures 6 and 10 show the spectral shapes of the input F-waves that allow for the generation of SH waves but not the spectral shape of the output SH-beam (or pulse). In order to accurately estimate the pulse broadening (or distortion) after passing through the crystal, the parameters for dispersion and other third-order nonlinearities at the resonances that we specified in this study (satisfying the BPM and zero GVM, simultaneously) should be first measured, but this has not been done enough yet in the previous publications. In other words, sufficient information has not yet been accumulated for pulse analysis after passing the crystal at resonances located in the spectral region where GVM is zero. We therefore believe that our study is very timely and can provide initial conditions for researchers to conduct potential experiments with light pulses in a simultaneous BPM–GV matching scheme. We hope that this study will inspire our readers and help them in their future experimental research using optical pulses.

4. Conclusions

We have theoretically and numerically investigated the broadband SHG properties of non-oxide NLO crystals including uniaxial chalcopyrite semiconductors and defect chalcopyrite (i.e., AGS, AGSe, CSP, HGS, TASe, GaSe, CGA, ZGP, and CdSe) as well as biaxial orthorhombic ternary chalcogenides (i.e., LIS, LISe, LGS, and LGSe). The BPM and GV matching properties (including the spectral position of resonance and the corresponding range of the F-vector direction), effective nonlinearities, spatial walk-offs, and acceptable F-wave spectra and bandwidths were discussed both theoretically and numerically. The effective nonlinearities were generally high in the uniaxial birefringent crystals, whereas the biaxial birefringent crystals had the advantage of exhibiting angular tunability of the spectral resonance. The calculated acceptable bandwidths of the F-wave range from 488.69 nm to 1164.24 nm (for uniaxial crystals) or from 369.30 nm to 648.12 nm (for biaxial crystals) in FWHM. Such broadband SHG using short pulse trains can potentially be applied to frequency up-conversion imaging in the mid-IR region, in information transmission using optical pulse signals, and in nonlinear optical signal processing.

Author Contributions: Conceptualization, K.J.L.; simulation, D.L. and K.J.L.; writing—original draft preparation, I.K. and D.L.; writing—review and editing, K.J.L.; supervision, K.J.L. All authors have read and agreed to the published version of the manuscript.

Funding: This research was funded by the National Research Foundation of Korea (grant NRF-2019R1F1A1063937); Korea Institute of Science and Technology (grant 2E29580-19-147); and Institute for Information and Communications Technology Promotion (IITP), grant funded by the Korean government (MSIT) (number 2020-0-00947).

Institutional Review Board Statement: Not applicable.

Informed Consent Statement: Not applicable.

Data Availability Statement: The data presented in this study are available in this article.

Conflicts of Interest: The authors declare no conflict of interest.

References

1. Krier, A. *Mid-Infrared Semiconductor Optoelectronics*; Springer: London, UK, 2006; pp. 595–738.
2. Ebrahim-Zadeh, M.; Sorokina, I.T. *Mid-Infrared Coherent Sources and Applications*; Springer: Dordrecht, The Netherlands, 2008; pp. 467–621.
3. Pereira, M.F.; Shulika, O. *Terahertz and Mid Infrared Radiation*; Springer: Dordrecht, The Netherlands, 2011; pp. 1–194.
4. Schunemann, P.G.; Zawilski, K.T.; Pomeranz, L.A.; Creeden, D.J.; Budni, P.A. Advances in nonlinear optical crystals for mid-infrared coherent sources. *J. Opt. Soc. Am. B* **2016**, *33*, D36–D43. [CrossRef]
5. Luo, X.; Li, Z.; Guo, Y.; Yao, J.; Wu, Y. Recent progress on new infrared nonlinear optical materials with application prospect. *J. Solid State Chem.* **2019**, *270*, 674–687. [CrossRef]
6. O'Connor, M.V.; Watson, M.A.; Shepherd, D.P.; Hanna, D.C.; Price, J.H.V.; Malinowski, A.; Nilsson, J.; Broderick, N.G.R.; Richardson, D.J.; Lefort, L. Synchronously pumped optical parametric oscillator driven by a femtosecond mode-locked fiber laser. *Opt. Lett.* **2002**, *27*, 1052–1054. [CrossRef]
7. Kumar, S.C.; Samanta, G.K.; Ebrahim-Zadeh, M. High-power, single-frequency, continuous-wave second-harmonic-generation of ytterbium fiber laser in PPKTP and MgO:sPPLT. *Opt. Express* **2009**, *17*, 13711–13726. [CrossRef]
8. Kokabee, O.; Esteban-Martin, A.; Ebrahim-Zadeh, M. Efficient, high-power, ytterbium-fiber-laser-pumped picosecond optical parametric oscillator. *Opt. Lett.* **2010**, *35*, 3210–3212. [CrossRef]
9. Kumar, S.C.; Ebrahim-Zadeh, M. High-power, fiber-laser-pumped, picosecond optical parametric oscillator based on MgO:sPPLT. *Opt. Express* **2011**, *19*, 26660–26665. [CrossRef] [PubMed]
10. Andriukaitis, G.; Balčiūnas, T.; Ališauskas, S.; Pugžlys, A.; Baltuška, A.; Popmintchev, T.; Chen, M.-C.; Murnane, M.M.; Kapteyn, H.C. 90 GW peak power few-cycle mid-infrared pulses from an optical parametric amplifier. *Opt. Lett.* **2011**, *36*, 2755–2757. [CrossRef] [PubMed]
11. Petrov, V. Parametric down-conversion devices: The coverage of the mid-infrared spectral range by solid-state laser sources. *Opt. Mat.* **2012**, *34*, 536–554. [CrossRef]
12. Petrov, V. Frequency down-conversion of solid-state laser sources to the mid-infrared spectral range using non-oxide nonlinear crystals. *Prog. Quantum Electron.* **2015**, *42*, 1–106. [CrossRef]
13. Zahedpour, S.; Hancock, S.W.; Milchberg, H.M. Ultrashort infrared 2.5–11 µm pulses: Spatiotemporal profiles and absolute nonlinear response of air constituents. *Opt. Lett.* **2019**, *44*, 843–846. [CrossRef]
14. Ionin, A.A.; Kinyaevskiy, I.O.; Klimachev, Y.M.; Kotkov, A.A.; Kozlov, A.Y.; Sagitova, A.M.; Sinitsyn, D.V.; Rulev, O.A.; Badikov, V.V.; Badikov, D.V. Frequency conversion of mid-IR lasers into the long-wavelength domain of 12–20 µm with $AgGaSe_2$, $BaGa_2GeSe_6$ and $PbIn_6Te_{10}$ nonlinear crystals. *Opt. Express* **2019**, *27*, 24353–24361. [CrossRef]
15. Kumar, S.C.; Schunemann, P.G.; Zawilski, K.T.; Ebrahim-Zadeh, M. Advances in ultrafast optical parametric sources for the mid-infrared based on $CdSiP_2$. *J. Opt. Soc. Am. B* **2016**, *33*, D44–D56. [CrossRef]
16. Ferdinandus, M.R.; Gengler, J.J.; Averett, K.L.; Zawilski, K.T.; Schunemann, P.G.; Liebig, C.M. Nonlinear optical measurements of $CdSiP_2$ at near and mid-infrared wavelengths. *Opt. Mater. Express* **2020**, *10*, 2066–2074. [CrossRef]
17. Popien, S.; Beutler, M.; Rimke, I.; Badikov, D.; Badikov, V.; Petrov, V. Femtosecond Yb-fiber laser synchronously pumped $HgGa_2S_4$ optical parametric oscillator tunable in the 4.4- to 12-µm range. *Opt. Eng.* **2018**, *57*, 111802. [CrossRef]
18. Liu, G.-Y.; Chen, Y.; Yao, B.-Q.; Wang, R.; Yang, K.; Yang, C.; Mi, S.; Dai, T.-Y.; Duan, X.-M. 3.5 W long-wave infrared $ZnGeP_2$ optical parametric oscillator at 9.8 µm. *Opt. Lett.* **2020**, *45*, 2347–2350. [CrossRef]
19. Liu, G.; Mi, S.; Yang, K.; Wei, D.; Li, J.; Yao, B.; Yang, C.; Dai, T.; Duan, X.; Tian, L.; et al. 161 W middle infrared $ZnGeP_2$ MOPA system pumped by 300 W-class Ho:YAG MOPA system. *Opt. Lett.* **2021**, *46*, 82–85. [CrossRef]

20. Nam, S.-H.; Fedorov, V.; Mirov, S.; Hong, K.-H. Octave-spanning mid-infrared femtosecond OPA in a ZnGeP$_2$ pumped by a 2.4 µm Cr:ZnSe chirped-pulse amplifier. *Opt. Express* **2020**, *28*, 32403–32414. [CrossRef]
21. Qian, C.; Yu, T.; Liu, J.; Jiang, Y.; Wang, S.; Shi, X.; Ye, X.; Chen, W. A High-Energy, Narrow-Pulse-Width, Long-Wave Infrared Laser Based on ZGP Crystal. *Crystals* **2021**, *11*, 656. [CrossRef]
22. Chen, Y.; Liu, G.; Yang, C.; Yao, B.; Wang, R.; Mi, S.; Yang, K.; Dai, T.; Duan, X.; Ju, Y. 1 W, 10.1 µm, CdSe optical parametric oscillator with continuous-wave seed injection. *Opt. Lett.* **2020**, *45*, 2119–2122. [CrossRef]
23. Chen, Y.; Yang, C.; Liu, G.; Yao, B.; Wang, R.; Yang, K.; Mi, S.; Dai, T.; Duan, X.; Ju, Y. 11 µm, high beam quality idler-resonant CdSe optical parametric oscillator with continuous-wave injection-seeded at 2.58 µm. *Opt. Express* **2020**, *28*, 17056–17063. [CrossRef]
24. Wang, S.; Dai, S.; Jia, N.; Zong, N.; Li, C.; Shen, Y.; Yu, T.; Qiao, J.; Gao, Z.; Peng, Q.; et al. Tunable 7–12 µm picosecond optical parametric amplifier based on a LiInSe$_2$ mid-infrared crystal. *Opt. Lett.* **2017**, *42*, 2098–2101. [CrossRef]
25. Smetanin, S.N.; Jelínek, M.; Kubeček, V.; Kurus, A.F.; Vedenyapin, V.N.; Lobanov, S.I.; Isaenko, L.I. 50-µJ level, 20-picosecond, narrowband difference-frequency generation at 4.6, 5.4, 7.5, 9.2, and 10.8 µm in LiGaS$_2$ and LiGaSe$_2$ at Nd:YAG laser pumping and various crystalline Raman laser seedings. *Opt. Mater. Express* **2020**, *10*, 1881–1890. [CrossRef]
26. Chen, B.-H.; Wittmann, E.; Morimoto, Y.; Baum, P.; Riedle, E. Octave-spanning single-cycle middle-infrared generation through optical parametric amplification in LiGaS$_2$. *Opt. Express* **2019**, *27*, 21306–21318. [CrossRef] [PubMed]
27. Lee, K.J.; Liu, S.; Parmigiani, F.; Ibsen, M.; Petropoulos, P.; Gallo, K.; Richardson, D.J. OTDM to WDM format conversion based on quadratic cascading in a periodically poled lithium niobate waveguide. *Opt. Express* **2010**, *18*, 10282–10288. [CrossRef]
28. Barh, A.; Rodrigo, P.J.; Meng, L.; Pedersen, C.; Tidemand-Lichtenberg, P. Parametric upconversion imaging and its applications. *Adv. Opt. Photon.* **2019**, *11*, 952–1019. [CrossRef]
29. Tan, S.; Wei, X.; Li, B.; Lai, Q.T.K.; Tsia, K.K.; Wong, K.K.Y. Ultrafast optical imaging at 2.0 µm through second-harmonic-generation-based time-stretch at 1.0 µm. *Opt. Lett.* **2018**, *43*, 3822–3825. [CrossRef] [PubMed]
30. Petrov, V.; Ghotbi, M.; Kokabee, O.; Esteban-Martin, A.; Noack, F.; Gaydardzhiev, A.; Nikolov, I.; Tzankov, P.; Buchvarov, I.; Miyata, K.; et al. Femtosecond nonlinear frequency conversion based on BiB$_3$O$_6$. *Laser Photon. Rev.* **2010**, *4*, 53–98. [CrossRef]
31. Feigelson, R.S.; Route, R.K. Recent developments in the growth of chalcopyrite crystals for nonlinear infrared applications. *Opt. Eng.* **1987**, *26*, 113–119. [CrossRef]
32. Barnes, N.P.; Gettemy, D.J.; Hietanen, J.R.; Lannini, R.A. Parametric amplification in AgGaSe$_2$. *Appl. Opt.* **1989**, *28*, 5162–5168.
33. Zawilski, K.T.; Schunemann, P.G.; Pollak, T.C.; Zelmon, D.E.; Fernelius, N.C.; Hopkins, F.K. Growth and characterization of large CdSiP$_2$ single crystals. *J. Cryst. Growth* **2010**, *312*, 1127–1132. [CrossRef]
34. Schunemann, P.G.; Pollak, T.M. Synthesis and growth of HgGa$_2$S$_4$ crystals. *J. Cryst. Growth* **1997**, *174*, 278–282. [CrossRef]
35. Ewbank, M.D.; Newman, P.R.; Mota, N.L.; Lee, S.M.; Wolfe, W.L.; DeBell, A.G.; Harrison, W.A. The temperature dependence of optical and mechanical properties of Tl$_3$AsSe$_3$. *J. Appl. Phys.* **1980**, *51*, 3848–3852. [CrossRef]
36. Vodopyanov, K.L.; Kulevskii, L.A.; Voevodin, V.G.; Gribenyukov, A.I.; Allakhverdiev, K.R.; Kerimov, T.A. High efficiency middle IR parametric superradiance in ZnGeP2 and GaSe crystals pumped by an erbium laser. *Opt. Commun.* **1991**, *83*, 322–326. [CrossRef]
37. Byer, R.L.; Kildal, H.; Feigelson, R.S. CdGeAs$_2$—A new nonlinear crystal phasematchable at 10.6 µm. *Appl. Phys. Lett.* **1971**, *19*, 237–240. [CrossRef]
38. Finsterbusch, K.; Bayer, A.; Zacharias, H. Tunable, narrow-band picosecond radiation in the mid-infrared by difference frequency mixing in GaSe and CdSe. *Appl. Phys. B* **2004**, *79*, 457–462. [CrossRef]
39. Kato, K.; Umemura, N. Sellmeier and thermo-optic dispersion formulas for LiInS$_2$. *Appl. Opt.* **2014**, *53*, 7998–8001. [CrossRef]
40. Petrov, V.; Zondy, J.-J.; Bidault, O.; Isaenko, L.; Vedenyapin, V.; Yelisseyev, A.; Chen, W.; Tyazhev, A.; Lobanov, S.; Marchev, G.; et al. Optical, thermal, electrical, damage, and phase-matching properties of lithium selenoindate. *J. Opt. Soc. Am. B* **2010**, *27*, 1902–1927. [CrossRef]
41. Isaenko, L.; Yelisseyev, A.; Lobanov, S.; Titov, A.; Petrov, V.; Zondy, J.-J.; Krinitsin, P.; Merkulov, A.; Vedenyapin, V.; Smirnova, J. Growth and properties of LiGaX$_2$ (X = S, Se, Te) single crystals for nonlinear optical applications in the mid-IR. *Cryst. Res. Technol.* **2003**, *38*, 379–387. [CrossRef]
42. Boyd, R.W. *Nonlinear Optics*, 4th ed.; Academic Press: San Diego, CA, USA, 2020; pp. 72–79.
43. Kato, K.; Petrov, V.; Umemura, N. Sellmeier and thermo-optic dispersion formulas for LiInSe$_2$. *Appl. Opt.* **2014**, *53*, 1063–1066. [CrossRef]
44. Kato, K.; Miyata, K.; Isaenko, L.; Lobanov, S.; Vedenyapin, V.; Petrov, V. Phase-matching properties of LiGaS$_2$ in the 1.025–10.5910 µm spectral range. *Opt. Lett.* **2017**, *42*, 4363–4366. [CrossRef] [PubMed]
45. Miyata, K.; Petrov, V.; Kato, K. Phase-matching properties of LiGaSe$_2$ for SHG and SFG in the 1.026–10.5910 µm range. *Appl. Opt.* **2017**, *56*, 6126–6129. [CrossRef]
46. Yariv, A.; Yeh, P. *Photonics: Optical Electronics in Modern Communications*, 6th ed.; Oxford University Press: New York, NY, USA, 2007; pp. 30–33.
47. Midwinter, J.E.; Warner, J. The effects of phase matching method and of uniaxial crystal symmetry on the polar distribution of second-order non-linear optical polarization. *Brit. J. Appl. Phys.* **1965**, *16*, 1135–1142. [CrossRef]
48. Dmitriev, V.G.; Nikogosyan, D.N. Effective nonlinearity coefficients for three-wave interactions in biaxial crystals of mm2 point group symmetry. *Opt. Comm.* **1993**, *95*, 173–182. [CrossRef]

49. Gehr, R.J.; Kimmel, M.W.; Smith, A.V. Simultaneous spatial and temporal walk-off compensation in frequency-doubling femtosecond pulses in β-BaB$_2$O$_4$. *Opt. Lett.* **1998**, *23*, 1298–1300. [CrossRef] [PubMed]
50. Zondy, J.-J.; Touahri, D.; Acef, O. Absolute value of the d_{36} nonlinear coefficient of AgGaS$_2$: Prospect for a low-threshold doubly resonant oscillator-based 3:1 frequency divider. *J. Opt. Soc. Am. B* **1997**, *14*, 2481–2497. [CrossRef]
51. Roberts, D.A. Simplified characterization of uniaxial and biaxial nonlinear optical crystals: A plea for standardization of nomenclature and conventions. *IEEE J. Quantum Electron.* **1992**, *28*, 2057–2074. [CrossRef]
52. Kemlin, V.; Boulanger, B.; Petrov, V.; Segonds, P.; Ménaert, B.; Schunneman, P.G.; Zawilski, K.T. Nonlinear, dispersive, and phase-matching properties of the new chalcopyrite CdSiP$_2$ [Invited]. *Opt. Mater. Express* **2011**, *1*, 1292–1300. [CrossRef]
53. Rotermund, F.; Petrov, V. Mercury thiogallate mid-infrared femtosecond optical parametric generator pumped at 1.25 µm by a Cr: Forsterite regenerative amplifier. *Opt. Lett.* **2000**, *25*, 746–748. [CrossRef]
54. Feichtner, J.D.; Roland, G.W. Optical properties of a new nonlinear optical material: Tl$_3$AsSe$_3$. *Appl. Opt.* **1972**, *11*, 993–998. [CrossRef]
55. Tanaka, E.; Kato, K. Second-harmonic and sum-frequency generation in CdGeAs$_2$. In *MRS Symposium Proceedings*; McDaniel, D.L., Jr., Manasreh, M.O., Miles, R.H., Sivananthan, S., Eds.; Materials Research Society: Warrendale, PA, USA, 1998; Volume 484, pp. 475–479.
56. Kato, K. Second-harmonic and sum-frequency generation in ZnGeP$_2$. *Appl. Opt.* **1997**, *36*, 2506–2510. [CrossRef]
57. Fossier, S.; Salaün, S.; Mangin, J.; Bidault, O.; Thénot, I.; Zondy, J.-J.; Chen, W.; Rotermund, F.; Petrov, V.; Petrov, P.; et al. Optical, vibrational, thermal, electrical, damage, and phase-matching properties of lithium thioindate. *J. Opt. Soc. Am. B* **2004**, *21*, 1981–2007. [CrossRef]
58. Petrov, V.; Yelisseyev, A.; Isaenko, L.; Lobanov, S.; Titov, A.; Zondy, J.-J. Second harmonic generation and optical parametric amplification in the mid-IR with orthorhombic biaxial crystals LiGaS$_2$ and LiGaSe$_2$. *Appl. Phys. B* **2004**, *78*, 543–546. [CrossRef]
59. Kim, I.; Lee, D.; Lee, K.J. Study of type II SPDC in lithium niobate for high spectral purity photon pair generation. *Crystals* **2021**, *11*, 406. [CrossRef]
60. Brehat, F.; Wyncke, B. Calculation of double-refraction walk-off angle along the phase-matching directions in nonlinear biaxial crystals. *J. Phys. B At. Mol. Opt. Phys.* **1989**, *22*, 1891–1898. [CrossRef]
61. Lee, D.; Kim, I.; Lee, K.J. Investigation of 1064-nm pumped type II SPDC in potassium niobate for generation of high spectral purity photon pairs. *Crystals* **2021**, *11*, 599. [CrossRef]
62. Bai, Y.; Bandyopadhyay, N.; Tsao, S.; Slivken, S.; Razeghi, M. Room temperature quantum cascade lasers with 27% wall plug efficiency. *Appl. Phys. Lett.* **2011**, *98*, 181102. [CrossRef]
63. Lyakh, A.; Maulini, R.; Tsekoun, A.; Go, R.; Patel, C.K.N. Tapered 4.7 µm quantum cascade lasers with highly strained active region composition delivering over 4.5 watts of continuous wave optical power. *Opt. Express* **2012**, *20*, 4382–4388. [CrossRef]
64. Maulini, R.; Lyakh, A.; Tsekoun, A.; Kumar, C.; Patel, N. λ~7.1 µm quantum cascade lasers with 19% wallplug efficiency at room temperature. *Opt. Express* **2011**, *19*, 17203–17211. [CrossRef]
65. Lu, Q.Y.; Bai, Y.; Bandyopadhyay, N.; Slivken, S.; Razeghi, M. Room-temperature continuous wave operation of distributed feedback quantum cascade lasers with watt-level power output. *Appl. Phys. Lett.* **2010**, *97*, 231119. [CrossRef]
66. Lu, Q.Y.; Bai, Y.; Bandyopadhyay, N.; Slivken, S.; Razeghi, M. 2.4 W room temperature continuous wave operation of distributed feedback quantum cascade lasers. *Appl. Phys. Lett.* **2011**, *98*, 181106. [CrossRef]
67. Lee, B.G.; Zhang, H.A.; Pflueg, C.; Diehl, L.; Belkin, M.A.; Fischer, M.; Wittmann, A.; Faist, J.; Capasso, F. Broadband distributed-feedback quantum cascade laser array operating from 8.0 to 9.8 µm. *IEEE Photon. Technol. Lett.* **2009**, *21*, 914–916. [CrossRef]
68. Hugi, A.; Maulini, R.; Faist, J. External cavity quantum cascade laser. *Semicond. Sci. Technol.* **2010**, *25*, 083001. [CrossRef]
69. Maulini, R.; Beck, M.; Faist, J.; Gini, E. Broadband tuning of external cavity bound-to-continuum quantum-cascade lasers. *Appl. Phys. Lett.* **2004**, *84*, 1659. [CrossRef]
70. Fedorov, V.V.; Mirov, S.B.; Gallian, A.; Badikov, D.V.; Frolov, M.P.; Korostelin, Y.V.; Kozlovsky, V.I.; Landman, A.I.; Podmar'kov, Y.P.; Akimov, V.A.; et al. 3.77–5.05-µm tunable solid-state lasers based on Fe^{2+}-doped ZnSe crystals operating at low and room temperatures. *IEEE J. Quantum Electron.* **2006**, *42*, 907. [CrossRef]
71. Firsov, K.N.; Gavrishchuk, E.M.; Kazantsev, S.Y.; Kononov, I.G.; Rodin, S.A. Increasing the radiation energy of ZnSe:Fe^{2+} laser at room temperature. *Laser Phys. Lett.* **2014**, *11*, 085001. [CrossRef]
72. Velikanov, S.D.; Gavrishchuk, E.M.; Zaretsky, N.A.; Zakhryapa, A.V.; Ikonnikov, V.B.; Kazantsev, S.Y.; Kononov, I.G.; Maneshkin, A.A.; Mashkovskii, D.A.; Saltykov, E.V.; et al. Repetitively pulsed Fe:ZnSe laser with an average output power of 20 W at room temperature of the polycrystalline active element. *Quantum Electron.* **2017**, *47*, 303. [CrossRef]
73. Kozlovsky, V.I.; Akimov, V.A.; Frolov, M.P.; Korostelin, Y.V.; Landman, A.I.; Martovitsky, V.P.; Mislavskii, V.V.; Podmar'kov, Y.P.; Skasyrsky, Y.K.; Voronov, A.A. Room-temperature tunable midinfrared lasers on transition-metal doped II-VI compound crystals grown from vapor phase. *Phys. Status Solidi B* **2010**, *247*, 1553. [CrossRef]
74. Cui, Y.; Huang, W.; Wang, Z.; Wang, M.; Zhou, Z.; Li, Z.; Gao, S.; Wang, Y.; Wang, P. 4.3 µm fiber laser in CO$_2$-filled hollow-core silica fibers. *Optica* **2019**, *6*, 951–954. [CrossRef]
75. Wittmann, A.; Hugi, A.; Gini, E.; Hoyler, N.; Faist, J. Heterogeneous high-performance quantum-cascade laser sources for broad-band tuning. *IEEE J. Quantum Electron.* **2008**, *44*, 1083–1088. [CrossRef]
76. Takaoka, E.; Kato, K.; Umemura, N. Thermo-optic dispersion formula for AgGaS$_2$. *Appl. Opt.* **1999**, *38*, 4577–4580. [CrossRef]

77. Komine, H.; Fukumoto, J.M.; Long, W.H.; Stappaerts, E.A. Noncritically phase matched mid-infrared generation in AgGaSe$_2$. *IEEE J. Sel. Top. Quant. Electron.* **1995**, *1*, 44–49. [CrossRef]
78. Kato, K.; Petrov, V.; Umemura, N. Phase-matching properties of yellow color HgGa$_2$S$_4$ for SHG and SFG in the 0.944–10.5910 μm range. *Appl. Opt.* **2016**, *55*, 3145–3148. [CrossRef] [PubMed]
79. Takaoka, E.; Kato, K. Temperature phase-matching properties for harmonic generation in GaSe. *Jpn. J. Appl. Phys.* **1999**, *38*, 2755–2759. [CrossRef]
80. Zelmon, D.E.; Hanning, E.A.; Schunemann, P.G. Refractive-index measurements and Sellmeier coefficients for zinc germanium phosphide from 2 to 9 μm with implications for phase matching in optical frequency-conversion devices. *J. Opt. Soc. Am. B* **2001**, *18*, 1307–1310. [CrossRef]
81. Bhar, G.C. Refractive index interpolation in phase-matching. *Appl. Opt.* **1976**, *15*, 305–307. [CrossRef] [PubMed]
82. Liu, P.Q.; Hoffman, A.J.; Escarra, M.D.; Franz, K.J.; Khurgin, J.B.; Dikmelik, Y.; Wang, X.; Fan, J.-Y.; Gmachl, C.F. Highly power-efficient quantum cascade lasers. *Nat. Photon.* **2010**, *4*, 95. [CrossRef]
83. Bai, Y.; Slivken, S.; Kuboya, S.; Darvish, S.R.; Razeghi, M. Quantum cascade lasers that emit more light than heat. *Nat. Photon.* **2010**, *4*, 99. [CrossRef]
84. Cathabard, O.; Teissier, R.; Devenson, J.; Baranov, A.N. InAs-based distributed feedback quantum cascade lasers. *Electron. Lett.* **2009**, *45*, 1028. [CrossRef]
85. M Squared Lasers Limited. Available online: https://www.m2lasers.com/firefly-ir.html (accessed on 7 July 2021).
86. Fedorov, V.Y.; Tzortzakis, S. Optimal wavelength for two-color filamentation-induced terahertz sources. *Opt. Express* **2018**, *26*, 31150–31159. [CrossRef]
87. Koulouklidis, A.D.; Gollner, C.; Shumakova, V.; Fedorov, V.Y.; Pugžlys, A.; Baltuška, A.; Tzortzakis, S. Observation of extremely efficient terahertz generation from mid-infrared two-color laser filaments. *Nat. Commun.* **2020**, *11*, 292. [CrossRef]

Article

Mid-IR Optical Property of Dy:CaF$_2$-SrF$_2$ Crystal Fabricated by Multicrucible Temperature Gradient Technology

Lihe Zheng [1], Jianbin Zhao [1], Yangxiao Wang [2], Weichao Chen [3,*], Fangfang Ruan [4,*], Hui Lin [5], Yanyan Xue [6], Jian Liu [6], Yang Liu [1], Ruiqin Yang [1], Haifeng Lu [1], Xiaodong Xu [7] and Liangbi Su [2]

1. Key Laboratory of Yunnan Provincial Higher Education Institutions for Optoelectronics Device Engineering, School of Physics and Astronomy, Yunnan University, Kunming 650500, China; zhenglihe@ynu.edu.cn (L.Z.); jianbinzhao@mail.ynu.edu.cn (J.Z.); liuyang1@mail.ynu.edu.cn (Y.L.); yangruiqin@mail.ynu.edu.cn (R.Y.); 2943277027@mail.ynu.edu.cn (H.L.)
2. Key Laboratory of Transparent Opto-functional Inorganic Materials, Shanghai Institute of Ceramics, Chinese Academy of Sciences, Shanghai 201899, China; wangtim@student.sic.ac.cn (Y.W.); suliangbi@sic.ac.cn (L.S.)
3. School of Technology, Pu'er University, Pu'er 665000, China
4. Department of Medical Imaging, Hangzhou Medical College, Hangzhou 310053, China
5. Engineering Research Center of Optical Instrument and System, Ministry of Education and Shanghai Key Lab of Modern Optical System, University of Shanghai for Science and Technology, Shanghai 200093, China; linh8112@163.com
6. School of Physics Science and Engineering, Tongji University, Shanghai 200092, China; xueyanyanf@163.com (Y.X.); 1910102@tongji.edu.cn (J.L.)
7. Jiangsu Key Laboratory of Advanced Laser Materials and Devices, School of Physics and Electronic Engineering, Jiangsu Normal University, Xuzhou 221116, China; xdxu79@jsnu.edu.cn
* Correspondence: weichaochen2010@foxmail.com (W.C.); ruan.f@hmc.edu.cn (F.R.)

Abstract: Dy^{3+}-doped CaF$_2$-SrF$_2$ crystals with various Dy^{3+} dopant concentrations were synthesized by multicrucible temperature gradient technology (MC-TGT). Dy:CaF$_2$-SrF$_2$ crystals were fluorite structured and crystallized in cubic $Fm\overline{3}m$ space group, as characterized by X-ray diffraction. The crystallographic site concentration was calculated from the measured density by Archimedes' hydrostatic weighing principle. The optical transmission reached over 90% with a sample thickness of 1.0 mm. The Sellmeier dispersion formula was obtained following the measured refractive index in a mid-IR range of 1.7–11 µm. Absorption coefficients of 6.06 cm^{-1} and 12.71 cm^{-1} were obtained at 804 nm and 1094 nm in 15% Dy:CaF$_2$-SrF$_2$ crystal. The fluorescence spectra of 15 at.% Dy:CaF$_2$-SrF$_2$ showed the strongest wavelength peak at 2919 nm with a full width at half maximum (FWHM) of 267 nm under an excitation wavelength of 808 nm. The fluorescence lifetimes were illustrated for different Dy^{3+} dopant levels of 5%, 10% and 15%. The results indicate that the Dy:CaF$_2$-SrF$_2$ crystal is a promising candidate for compact mid-IR lasers.

Keywords: Dy:CaF$_2$-SrF$_2$; crystal growth; temperature gradient technology; midinfrared crystal; Sellmeier dispersion formula

1. Introduction

Midinfrared (mid-IR) lasers are considered of great importance due to their wide applications in fundamental and practical fields such as directional infrared countermeasures, atmospheric monitoring, biomedicine, medical laser, optical communication and high-energy physics [1]. Directly pumped mid-IR solid-state lasers have attracted significant attention due to their advantages such as simple system composition, compact size, high efficiency and high output power. Active ions and host materials are both important in obtaining diode laser pumped solid-state lasers.

Rare-earth ions (Tm^{3+}, Er^{3+}, Ho^{3+} and Dy^{3+}) are the preferred active ions for laser emission in the mid-IR spectral range. Tm^{3+}, Er^{3+}, Ho^{3+} ions have been reported with 2–3 µm lasing in crystalline or glass hosts such as Tm:YAG [2], Tm:YLF [3], Tm:ZBLAN [4],

Tm:SrF$_2$ [5], Ho:CaF$_2$ [6], Ho:YAG [7], Er:SrF$_2$ [8] and Er:YSGG [9]. With energy transitions corresponding to 2.3–3.4 µm and 4.0–6.2 µm, Dy^{3+} is recognized as an active ion with high potential applications for mid-IR lasers. However, compared with other rare-earth ions, such as Tm^{3+}, Er^{3+} and Ho^{3+}, laser emissions from Dy^{3+} ion-based solid-state systems are relatively rare. In addition, the intrinsic emission wavelengths of Dy^{3+} include emission band peaks around 2.4, 3.4, 4.3 and 5.4 µm. Owing to the even electron number of 4f electron shells in Dy^{3+} and Stark effects in different crystal fields, a smooth broadband emission spectrum around 2.3–3.4 µm could lead to tunable or ultrafast laser output.

Recently, Dy^{3+} laser has been investigated in fluorozirconate ZBLAN glass fiber with compositions of ZrF$_4$, BaF$_2$, LaF$_3$, AlF$_3$ and NaF. In 2003, Jackson reported a 2.9 µm CW laser from Dy:ZBLAN with maximum output power of 275 mW and slope efficiency of 4.5% [10]. In 2011, Tsang and El-Taher demonstrated an efficient Dy:ZBLAN fiber lasing around 3 µm with slope efficiency of 23%. However, the output power was limited to 100 mW [11]. In 2016, Majewski et al. reported a Dy:ZBLAN laser at 3.04 µm, with a slope efficiency of 51% and maximum output of 80 mW [12]. In 2018, Woodward et al. successfully obtained watt-level Dy:ZBLAN fiber laser at 3.15 µm with slope efficiency of 73% [13]. In contrast, Dy^{3+}-doped crystals are yet to be explored for mid-IR lasers. Currently, the Dy^{3+}-doped crystalline host is limited in fluoride and thiogallate crystals such as Dy:BaY$_2$F$_8$ [14], Dy:LYF [15], Dy:PbGa$_2$S$_4$ [16,17] and Dy:CaGa$_2$S$_4$ [18]. Due to superior thermal, mechanical and moisture-proof properties, the Dy^{3+}-doped crystalline host is expected to have good performance in the mid-IR spectral range compared with those in ZBLAN fiber.

In this work, a family of CaF$_2$-SrF$_2$ crystals doped with Dy(III) was prepared to generate efficient mid-IR emission properties for potential applications in mid-IR lasers. CaF$_2$-SrF$_2$ crystal was selected as host material with a molar ratio of 1:1 for Ca/Sr, as inspired by predominant optical properties when doped with rare-earth ions such as Yb^{3+} [19,20]. Meanwhile, CaF$_2$-SrF$_2$ possesses low phonon energy, which is of benefit to weaken nonradiative decay from intermediate states to lower ground states in rare-earth ions [21,22]. Moreover, CaF$_2$-SrF$_2$ is an azeotrope system, and even the composition ratio of Ca/Sr is different [23,24]. Karimov et al. reported the growth of mixed crystals at the azeotrope point [25].

This research work focuses on crystal growth of Dy:CaF$_2$-SrF$_2$ with different Dy^{3+} dopant concentrations. The refractive index and related Sellmeier dispersion formula were obtained in the range of 1.7–11 µm. Optical characterization was conducted in the mid-IR spectral range, followed by the discussions of the energy transfer path in Dy:CaF$_2$-SrF$_2$ crystal.

2. Materials and Methods

2.1. Crystal Growth

Dy:CaF$_2$-SrF$_2$ crystal boules, with Dy^{3+} concentrations of 5 at.%, 10 at.% and 15 at.%, were fabricated by multicrucible temperature gradient technology (MC-TGT). TGT is a directional solidification technique adapted for the growth of high-temperature crystals [26]. The traditional TGT method allows for one-crucible crystal growth. To enhance growth efficiency, we developed an MC-TGT that allows obtaining multiple crystal boules in the one-growth process [22]. Six crucibles were fixed in the furnace. Benefiting from the azeotrope properties of CaF$_2$-SrF$_2$, no crystal seed was used. A stable temperature gradient was then built around the crucible to conduct latent heat and to promote crystallization. The starting materials were CaF$_2$, SrF$_2$ and DyF$_3$ powders with purity higher than 99.995%. The raw materials were mixed according to Ca/Sr molar ratio of 1:1 before placing it in a graphite crucible. Then, 1 wt.% PbF$_2$ was added to remove oxygen. The melting points of DyF$_3$, CaF$_2$ and SrF$_2$ are 1360 °C, 1420 °C and 1477 °C, respectively. The furnace was insulated for 3 h at 1530 °C while evacuating to 10^{-3} Pa. The cooling rate for crystal growth was set at 1.5 °C/h. After growth, the crucible was cooled to room temperature at a cooling rate of 20 °C/h. The Dy:CaF$_2$-SrF$_2$ crystals were then cut and processed for subsequent tests.

2.2. Characterizations

The segregation coefficient of Dy^{3+} ion in the CaF$_2$-SrF$_2$ host was detected by inductively coupled plasma atomic emission spectrometry (ICP-AES). The sample was cut from the initial part of the crystal and then ground into fine powder in an agate mortar. The solvent was a mixture of phosphoric acid and boric acid. After obtaining the weight percentage of Dy^{3+}, Ca^{2+} and Sr^{2+} in the solvent, the dopant level of Dy^{3+} was calculated. The segregation coefficient could thus be calculated by dividing the measured dopant value of Dy^{3+} by the theoretical value. The segregation coefficient was measured for the sample adjacent to the initial part of the as-grown crystal boule.

The structure of Dy:CaF$_2$-SrF$_2$ crystal was measured by a powder-X-ray diffractometer (P-XRD, RIGAKU TTRIII-18KW, Tokyo, Japan) using a Cu target at room temperature. The scan rate was 3°/min. The raw data from the P-XRD pattern were analyzed to fix the diffraction peak by comparing with the standard Power Diffraction File (PDF) data from Jade software. The diffractograms were gathered by Origin software.

The refractive index in the mid-IR spectral range was characterized using an infrared ellipsometer (J.A. Woollam IR-Vase II, Lincoln, NE, USA). The single crystal along the growth axis was cut with a diameter of 20 mm and a thickness of 1 mm. One surface was polished, while the other surface was kept rough for measurement. The measured data were then analyzed by nonlinear fitting by Origin software to obtain dispersion formula.

The optical quality of as-grown crystal boules was characterized by mid-IR transmission spectra with a sampling step of 2.8 nm (Bruker TENSOR27, Karlsruhe, Germany). Absorption spectra were measured with a UV/vis/NIR spectrophotometer (Varian Cary 5000, Palo Alto, CA, USA) using Xe light as a pump source. Fluorescence spectra were measured with an 808 nm pump source (Edinburg Instruments FLS1000, Livingston, UK) and an InSb detector. As a comparison, a 1320 nm pump source from a pulse generator (Thurlby Thandar Instruments TGP 110, Huntingdon, UK) was used for fluorescence spectra, emitting at 1200–3500 nm, recorded by a digital phosphor oscilloscope with sampling rate of 1.25 GS/s and frequency of 100 MHz (Tektronix TDS 3012C, Beaverton, OR, USA). The fluorescence lifetime measurement was carried out by a computer-controlled transient digitizer decay curve of emission under a pump wavelength of 808 nm. Single crystals with a thickness of 1 mm along the growth axis were polished and then used for the above-mentioned spectroscopic measurements. All measurements were performed at room temperature.

3. Results and Discussion

3.1. Crystal Structure and Optical Quality

The traditional TGT method allows for one-crucible crystal growth. To enhance growth efficiency, we developed an MC-TGT that allows obtaining multiple crystal boules in the one-growth process. Figure 1 shows the obtained Dy:CaF$_2$-SrF$_2$ crystal boules up to 20 mm in diameter and 68 mm in length. The as-grown yellow boules are homogeneous without bubbles. The dopants of Dy^{3+} ions are 5 at.%, 10 at.% and 15 at.%. The segregation coefficient of Dy^{3+} ion in 5 at.% Dy:CaF$_2$-SrF$_2$ and 15 at.% Dy:CaF$_2$-SrF$_2$ is 1.0, while that in 10 at.% Dy:CaF$_2$-SrF$_2$ is 0.96. This indicates the high solubility of Dy^{3+} ion in the CaF$_2$-SrF$_2$ host lattice.

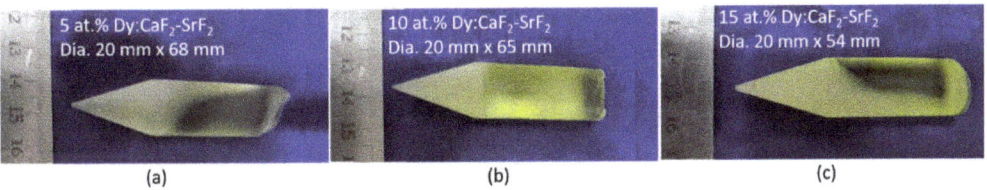

Figure 1. As-grown Dy:CaF$_2$-SrF$_2$ crystal boules using MC-TGT. (**a**) 5at.% Dy; (**b**) 10at.% Dy; (**c**) 15at.% Dy.

The method of Archimedes' hydrostatic weighing principle aids in the determination of density following Equation (1).

$$\rho = m_1 \cdot \rho_1 / (m_1 - m_2) \quad (1)$$

Here, m_1 and m_2 are sample weights measured in air and in water, respectively. ρ_1 is the density of water marked as 1 g·cm^{-3}. ρ is the sample density in air to be determined. The test results are listed in Table 1.

Table 1. Density determination following Archimedes' hydrostatic weighing principle.

Crystals	m_1 (g)	m_2 (g)	ρ (g/cm^3)
5% Dy:CaF$_2$-SrF$_2$	12.24010	9.20478	4.03256
10% Dy:CaF$_2$-SrF$_2$	14.00670	10.69937	4.23505
15% Dy:CaF$_2$-SrF$_2$	13.22491	10.28247	4.49454

Figure 2 gives the measured XRD pattern of Dy:CaF$_2$-SrF$_2$ crystal compared with that of pure CaF$_2$ referring to PDF Number 75-0363 and that of pure SrF$_2$ referring to PDF Number 06-0262. The main three strongest peaks in CaF$_2$ are located at 2θ of 28.272° for (111), 47.008° for (200) and 55.764° for (311), while those in SrF$_2$ are located at 2θ of 26.57° for (111), 44.123° for (200) and 52.273° for (311). In the case of Dy:CaF$_2$-SrF$_2$ crystal, the three strongest peaks are located at 2θ of 27.46°, 45.52° and 53.96°. The diffraction angles 2θ of Dy:CaF$_2$-SrF$_2$ situates in-between those of CaF$_2$ and SrF$_2$. This indicates that Dy:CaF$_2$-SrF$_2$ crystal is fluorite structured and crystallizes in the cubic $Fm\overline{3}m$ space group.

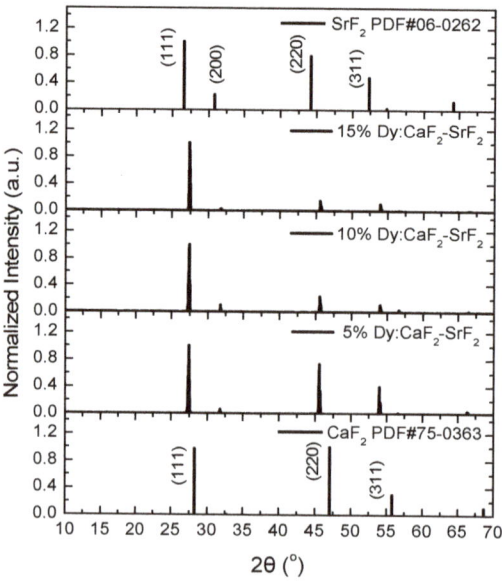

Figure 2. XRD pattern of Dy:CaF$_2$-SrF$_2$ crystal.

The optical quality of polished Dy:CaF$_2$-SrF$_2$ crystals was characterized by transmission spectra. Figure 3 gives the transmission spectra of Dy:CaF$_2$-SrF$_2$ at 2–11 μm. The transmission curves in Figure 3 are based on measured raw data subtracting the background without considering the reflections on both surfaces. It shows that the transmission of all three polished samples is above 90% from 4 to 9 μm. The infrared transmittance cut-off wavelength is around 11 μm. It could be concluded that the crystallinity and optical quality of Dy:CaF$_2$-SrF$_2$ crystal boules are good.

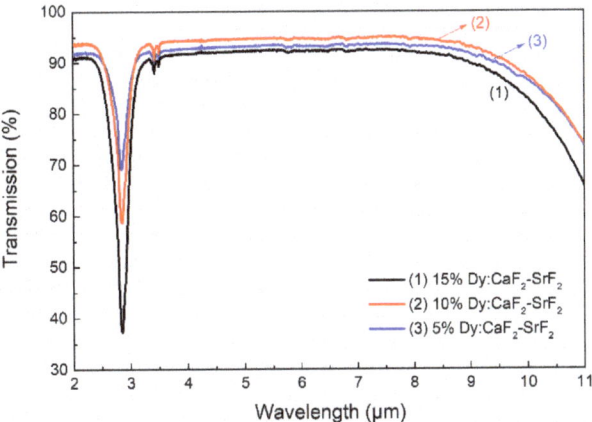

Figure 3. Transmission spectra of Dy:CaF$_2$-SrF$_2$ crystals at 2–11 μm.

3.2. Refractive Index in the Mid-IR Spectral Range

Figure 4 gives the refractive index of Dy:CaF$_2$-SrF$_2$ in the mid-IR spectral range of 1.7–11 μm. As seen from Figure 4, the refractive index of Dy:CaF$_2$-SrF$_2$ increases along with a higher dopant level of Dy^{3+}. The Sellmeier dispersion formula is used for nonlinear fitting. The dispersion formula for Dy:CaF$_2$-SrF$_2$ with Dy^{3+} dopant levels of 5%, 10% and 15% is thus achieved and described in Equations (2)–(4), respectively. The reduced Chi-square is 2.70×10^{-5}. The adjusted R^2 is 0.9986.

$$n^2 - 1 = 0.20292 + \frac{0.69272\lambda^2}{\lambda^2 - 0.94855^2} + \frac{0.87102\lambda^2}{\lambda^2 - 60.10^2} + \frac{5.64914\lambda^2}{\lambda^2 - 45.04^2} \quad (2)$$

$$n^2 - 1 = 0.21687 + \frac{0.70758\lambda^2}{\lambda^2 - 0.97111^2} + \frac{1.84394\lambda^2}{\lambda^2 - 70.44^2} + \frac{6.11963\lambda^2}{\lambda^2 - 48.55^2} \quad (3)$$

$$n^2 - 1 = 0.20292 + \frac{0.73684\lambda^2}{\lambda^2 - 0.97075^2} + \frac{2.30876\lambda^2}{\lambda^2 - 69.93^2} + \frac{6.15892\lambda^2}{\lambda^2 - 50.30^2} \quad (4)$$

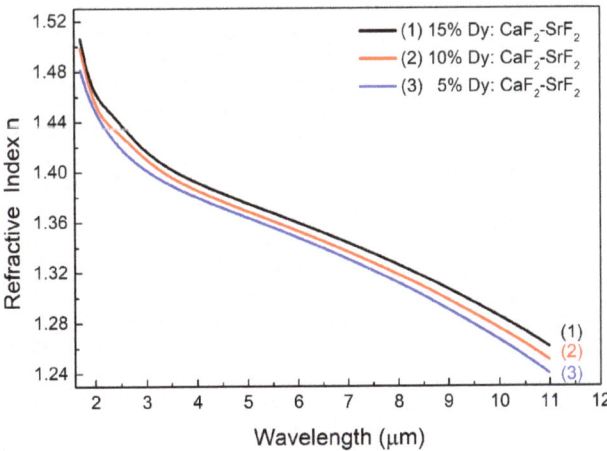

Figure 4. Refractive index of Dy:CaF$_2$-SrF$_2$ in the mid-IR spectral range of 1.7–11 μm.

3.3. Absorption and Emission Spectra

Figure 5 shows the room-temperature absorption spectra in the spectral range of 650–1900 nm and the assignment of energy level in Dy:CaF$_2$-SrF$_2$ crystals. The assignment of energy level could refer to that in Dy:BaY$_2$F$_8$ [27]. In the mid-IR region, Dy^{3+} ions show broad absorption bands in 5% Dy:CaF$_2$-SrF$_2$ crystal peaks at 804 nm, 907 nm, 1092 nm, 1276 nm and 1714 nm corresponding to absorption coefficients of 2.12 cm^{-1}, 2.69 cm^{-1}, 4.04 cm^{-1}, 2.54 cm^{-1} and 1.18 cm^{-1}, respectively. In the case of 10% Dy:CaF$_2$-SrF$_2$ crystal, the absorption coefficients are 3.56 cm^{-1}, 4.46 cm^{-1}, 7.08 cm^{-1}, 4.47 cm^{-1} and 2.02 cm^{-1}, corresponding to absorption band peaks at 804 nm, 908 nm, 1093 nm, 1273 nm and 1716 nm. For 15% Dy:CaF$_2$-SrF$_2$ crystal, it gives the strongest absorption coefficients of 6.06 cm^{-1}, 7.56 cm^{-1}, 12.71 cm^{-1}, 8.25 cm^{-1} and 3.63 cm^{-1}, corresponding to absorption band peaks at 804 nm, 907 nm, 1094 nm, 1277 nm and 1716 nm. The absorption band peaks at 1308 nm and 1720 nm correspond to energy level transitions of $^6H_{15/2} \rightarrow {}^6H_{9/2}$, $^6F_{11/2}$ and $^6H_{15/2} \rightarrow {}^6H_{11/2}$ in Dy^{3+} ion. With the increase in Dy^{3+} ion dopant levels from 5% to 15%, the absorption coefficient becomes 2.8–3.3 times stronger. Broad absorption bands are profitable in increasing the diode-pumping efficiency, as laser diodes typically emit in a narrow spectral range and present a thermal shift in the peak wavelength.

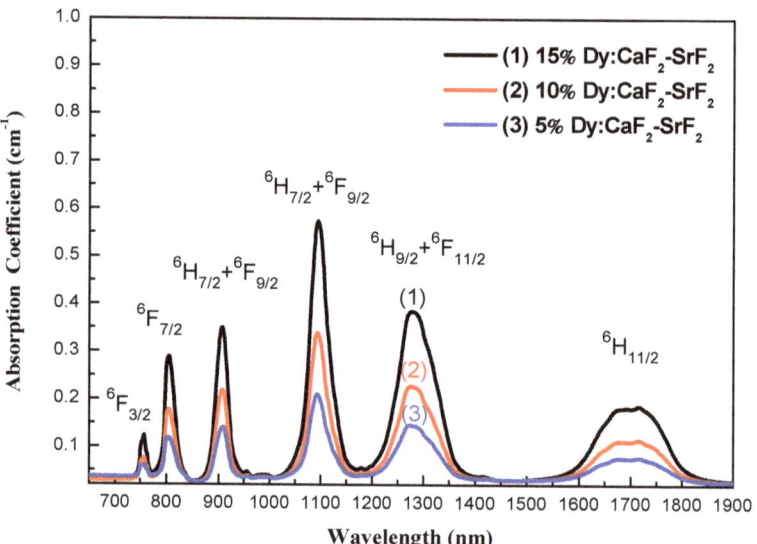

Figure 5. Absorption coefficient of Dy:CaF$_2$-SrF$_2$ crystal at 650–1900 nm.

Table 2 gives the calculated absorption cross-section of Dy^{3+} according to the expression $\sigma_{abs} = \alpha/N$. Here, α is the absorption coefficient of Dy^{3+}. N is the concentration of Dy^{3+} ions with 1.21 × 10^{21} ion·cm^{-3}, 2.42 × 10^{21} ion·cm^{-3} and 4.03 × 10^{21} ion·cm^{-3} in 5% Dy:CaF$_2$-SrF$_2$, 10% Dy:CaF$_2$-SrF$_2$ and 15% Dy:CaF$_2$-SrF$_2$, respectively.

Table 2. Absorption cross-section (σ_{abs}) of Dy^{3+} ions at 650–1900 nm.

Crystal	σ_{abs} (× 10^{-20} cm^2)			
	804 nm	907 nm	1094 nm	1287 nm
5% Dy:CaF$_2$-SrF$_2$	0.176	0.223	0.335	0.211
10% Dy:CaF$_2$-SrF$_2$	0.147	0.184	0.293	0.185
15% Dy:CaF$_2$-SrF$_2$	0.150	0.187	0.315	0.205

Figure 6 shows the emission spectra and peak assignment of Dy:CaF$_2$-SrF$_2$ crystal in the mid-IR range under excitation wavelengths of 808 nm and 1320 nm. In the case of using a pump wavelength of 1320 nm, as shown in Figure 6a, the fluorescence band peak at 1970 nm, corresponding to the energy level transition from $^6H_{9/2} + ^6F_{11/2}$ to $^6H_{15/2}$ [16], was detected with an intensity variation in the order of the Dy^{3+} doping concentration. The strongest fluorescence intensity appears in 5% Dy:CaF$_2$-SrF$_2$, while the lowest fluorescence intensity appears in 15% Dy:CaF$_2$-SrF$_2$. It is interesting to note that the intensity of the fluorescence band peak at 2882 nm is not affected by the Dy^{3+} doping concentration.

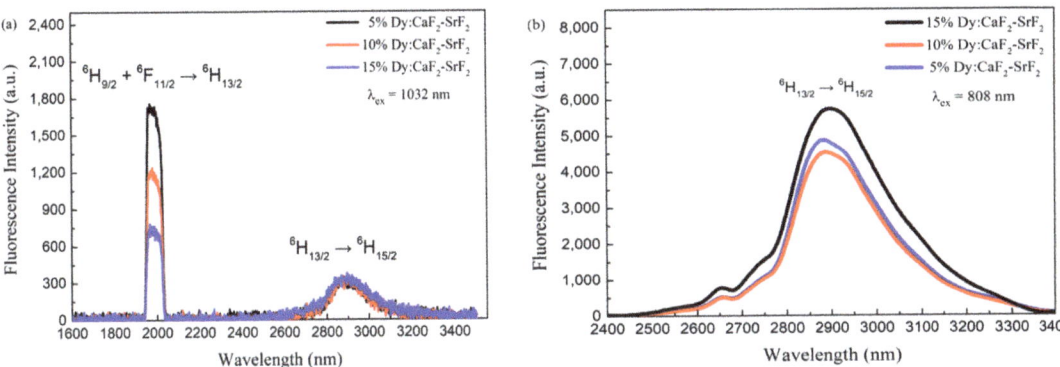

Figure 6. Emission spectra of Dy:CaF$_2$-SrF$_2$ crystal under different pump source. (**a**) 1320 nm pump; (**b**) 808 nm pump.

Figure 6b shows the emission spectra by using an excitation wavelength of 808 nm. The emission intensity of 10% Dy:CaF$_2$-SrF$_2$ is the lowest, while that of 15% Dy:CaF$_2$-SrF$_2$ is the highest. In order to illustrate the mid-IR fluorescence bands, nonlinear fittings for multiple peaks are used in the form of the Lorentz function. Taking 15% Dy:CaF$_2$-SrF$_2$ for example, the full width at half maximum (FWHM) is 267 nm for the emission band peak at 2919 nm, corresponding to the energy level transition from $^6H_{13/2}$ to $^6H_{15/2}$ of Dy^{3+} ion. In the case of Dy:CaF$_2$-SrF$_2$ with Dy^{3+} dopant levels of 10% and 5%, the values of FWHM are both 237 nm. All emission bands peak at 2913 nm.

Figure 7 gives the measured fluorescence lifetime curves for the energy transfer channel $^6H_{13/2} \rightarrow ^6H_{15/2}$ of Dy^{3+} ions in Dy:CaF$_2$-SrF$_2$ crystals. The obtained fluorescence lifetime curves are processed with nonlinear fitting according to the second-order exponential formula (ExpDec2), as described in Equation (5).

$$y = A_0 + A_1 e^{-\frac{\tau}{t_1}} + A_2 e^{-\frac{\tau}{t_2}} \quad (5)$$

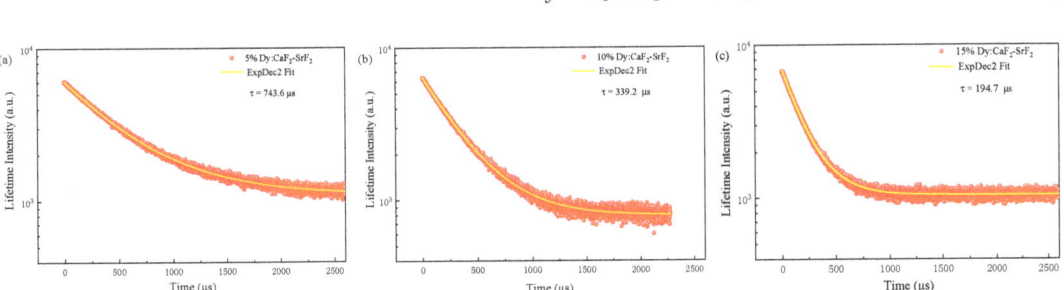

Figure 7. Measured and nonlinear fitted fluorescence lifetime curves in Dy:CaF$_2$-SrF$_2$. (**a**) 5% Dy; (**b**) 10% Dy; (**c**) 15% Dy.

Here, A_0, A_1, A_2, t_1 and t_2 are constant values that could be obtained from the fitting results. Accordingly, the second-order exponential formula for Dy:CaF$_2$-SrF$_2$ with various

Dy^{3+} dopant levels is thus written in Equations (6)–(8). The average fluorescence lifetime could be obtained from Equation (9). The fluorescence lifetimes are calculated as 743.6 µs, 339.2 µs and 194.7 µs for Dy:CaF$_2$-SrF$_2$ with Dy^{3+} dopant levels of 5%, 10% and 15%, respectively. The reduction of fluorescence lifetime along with a higher Dy^{3+} concentration could be attributed to the increase in interstitial fluoride ions. The mechanism could be described in the defect reaction equation as shown in Equation (10). It would be interesting to investigate the mechanism of enhancing fluorescence intensity while maintaining lifetime in future work.

$$y = 1046.495 + 331.646e^{-\frac{T}{1723.164}} + 4722.538e^{-\frac{T}{512.166}} \tag{6}$$

$$y = 793.798 + 5605.837e^{-\frac{T}{339.964}} + 60.970e^{-\frac{T}{104.241}} \tag{7}$$

$$y = 1050.882 + 2892.258e^{-\frac{T}{194.744}} + 2892.258e^{-\frac{T}{194.744}} \tag{8}$$

$$\tau = \frac{A_1 t_1^2 + A_2 t_2^2}{A_1 t_1 + A_2 t_2} \tag{9}$$

$$DyF_3 \xrightarrow{CaF_2-SrF_2} Dy_{Ca/Sr} + 2F_F + F'_i \tag{10}$$

4. Summary

This paper concerns the important issue of novel emissive materials for laser applications in the significant mid-IR range. Dy^{3+}-doped CaF$_2$-SrF$_2$ crystals with fluorite structure and cubic $Fm\bar{3}m$ space group were synthesized by MC-TGT. The crystallographic site concentrations were up to 4.034×10^{21} ions·cm^{-3} in 15 at.% Dy:CaF$_2$-SrF$_2$. The optical transmission of Dy:CaF$_2$-SrF$_2$ crystal reached over 90% with a sample thickness of 1.0 mm. The Sellmeier dispersion formula for 1.7–11 µm was obtained from the refractive index. The strongest absorption coefficients of 6.06 cm^{-1}, 7.56 cm^{-1}, 12.71 cm^{-1}, 8.25 cm^{-1} and 3.63 cm^{-1} were obtained in 15% Dy:CaF$_2$-SrF$_2$ crystal corresponding to absorption band peaks at 804 nm, 907 nm, 1094 nm, 1277 nm and 1716 nm. The value of FWHM was 267 nm for the emission band peak at 2919 nm in 15% Dy:CaF$_2$-SrF$_2$ under an excitation wavelength of 808 nm. The fluorescence lifetime in 5% Dy:CaF$_2$-SrF$_2$ was 3.8 times longer than that in 15% Dy:CaF$_2$-SrF$_2$. Further research work will focus on the enhancement of lifetime while maintaining strong emission in the mid-IR spectral range.

Author Contributions: Methodology—W.C.; formal analysis—J.Z., H.L. (Hui Lin), Y.X., Y.L. and R.Y.; investigation—H.L. (Haifeng Lu), Y.W., H.L. (Hui Lin), R.Y., J.L. and Y.X.; validation—F.R. and X.X.; funding acquisition—L.S.; supervision—L.Z.; writing—original draft preparation—W.C. and J.Z.; writing—review and editing—L.Z., F.R. and L.S. All authors have read and agreed to the published version of the manuscript.

Funding: This research was funded by National Natural Science Foundation of China, Grant Number U1830104; Opening Project of the State Key Laboratory of Transparent Opto-functional Inorganic Materials, Chinese Academy of Science, Grant Number KLTOIM202001; Yunnan Fundamental Research Projects, Grant Number 202101AT070162; Yunnan University First-class University Construction Project, Grant Number C176220100155; Innovative entrepreneurial training for Yunnan College Students, Grant Number 202010673075.

Data Availability Statement: Not applicable.

Acknowledgments: The authors acknowledge the experimental support from Shaohua Wu and Huajin Wang.

Conflicts of Interest: The authors declare no conflict of interest.

References

1. Xue, Y.Y.; Xu, X.D.; Su, L.B.; Xu, J. Research Progress of Mid-infrared Laser Crystals. *J. Synth. Cryst.* **2020**, *49*, 1347–1360.
2. Cao, D.; Peng, Q.; Du, S.; Xu, J.; Guo, Y.; Yang, J.; Bo, Y.; Zhang, J.; Cui, D.; Xu, Z. A 200 W diode-side-pumped CW 2 µm Tm:YAG laser with water cooling at 8 °C. *Appl. Phys. B* **2011**, *103*, 83–88. [CrossRef]

3. Wang, S.Q.; Huang, H.T.; Chen, H.W.; Liu, X.; Liu, S.D.; Xu, J.L.; Shen, D.Y. High efficiency nanosecond passively Q-switched 2.3 μm Tm:YLF laser using a ReSe$_2$-based saturable output coupler. *OSA Contin.* **2019**, *2*, 1676–1682. [CrossRef]
4. Lancaster, D.G.; Gross, S.; Ebendorff-Heidepriem, H.; Fuerbach, A.; Withford, M.J.; Monro, T.M. 2.1 μm waveguide laser fabricated by femtosecond laser direct-writing in Ho^{3+}, Tm^{3+}:ZBLAN glass. *Opt. Lett.* **2012**, *37*, 996–998. [CrossRef] [PubMed]
5. Sottile, A.; Damiano, E.; Rabe, M.; Bertram, R.; Klimm, D.; Tonelli, M. Widely tunable, efficient 2 μm laser in monocrystalline Tm^{3+}:SrF$_2$. *Opt. Express* **2018**, *26*, 5368–5380. [CrossRef] [PubMed]
6. Duan, X.M.; Shen, Y.J.; Zhang, Z.; Su, L.B.; Dai, T.Y. A passively Q-switching of diode-pumped 2.08 μm Ho:CaF$_2$ laser. *Infrared Phys. Technol.* **2019**, *103*, 103071. [CrossRef]
7. Duan, X.M.; Shen, Y.J.; Yao, B.Q.; Wang, Y.Z. A 106 W Q-switched Ho:YAG laser with single crystal. *Optik* **2018**, *169*, 224–227. [CrossRef]
8. Fan, M.Q.; Li, T.; Zhao, J.; Zhao, S.Z.; Li, G.Q.; Yang, K.J.; Su, L.B.; Ma, H.Y.; Kränkel, C. Continuous wave and ReS$_2$ passively Q-switched Er:SrF$_2$ laser at ~3 μm. *Opt. Lett.* **2018**, *43*, 1726–1729. [CrossRef]
9. Shen, B.J.; Kang, H.X.; Chen, P.; Liang, J.; Ma, Q.; Fang, J.; Sun, D.L.; Zhang, Q.L.; Yin, S.T.; Yan, X.P.; et al. Performance of continuous-wave laser-diode side-pumped Er:YSGG slab lasers at 2.79 μm. *Appl. Phys. B* **2015**, *121*, 511–515. [CrossRef]
10. Jackson, S.D. Continuous wave 2.9 μm dysprosium-doped fluoride fiber laser. *Appl. Phys. Lett.* **2003**, *83*, 1316–1318. [CrossRef]
11. Tsang, Y.H.; El-Taher, A.E. Efficient lasing at near 3 μm by a Dy-doped ZBLAN fiber laser pumped at ~1.1 μm by an Yb fiber laser. *Laser Phys. Lett.* **2011**, *8*, 818–822. [CrossRef]
12. Majewski, M.R.; Jackson, S.D. Highly efficient mid-infrared dysprosium fiber laser. *Opt. Lett.* **2016**, *41*, 2173–2176. [CrossRef] [PubMed]
13. Woodward, R.I.; Majewski, M.R.; Bharathan, G.; Hudson, D.D.; Fuerbach, A.; Jackson, S.D. Watt-level dysprosium fiber laser at 3.15 μm with 73% slope efficiency. *Opt. Lett.* **2018**, *43*, 1471–1474. [CrossRef] [PubMed]
14. Djeu, N.; Hartwell, V.E.; Kaminskii, A.A.; Butashin, A.V. Room-temperature 3.4-μm Dy:BaYb$_2$F$_8$ laser. *Opt. Lett.* **1997**, *22*, 997–999. [CrossRef]
15. Barnes, N.P.; Allen, R.E. Room temperature Dy:YLF laser operation at 4.34 μm. *IEEE J. Quantum Electron.* **1991**, *27*, 277–282. [CrossRef]
16. Jelinkova, H.; Doroshenko, M.E.; Osiko, V.V.; Jelínek, M.; Šulc, J.; Němec, M.; Vyhlídal, D.; Badikov, V.V.; Badikov, D.V. Dysprosium thiogallate laser: Source of mid-infrared radiation at 2.4, 4.3, and 5.4 μm. *Appl. Phys. A* **2016**, *122*, 738. [CrossRef]
17. Jelínková, H.; Doroshenko, M.E.; Jelínek, M.; Šulc, J.; Osiko, V.V.; Badikov, V.V.; Badikov, D.V. Dysprosium-doped PbGa$_2$S$_4$ laser generating at 4.3 μm directly pumped by 1.7 μm laser diode. *Opt. Lett.* **2013**, *38*, 3040–3043. [CrossRef]
18. Nostrand, M.C.; Page, R.H.; Payne, S.A.; Krupke, W.F.; Schunemann, P.G. Room-temperature laser action at 4.3–4.4 μm in CaGa$_2$S$_4$:Dy^{3+}. *Opt. Lett.* **1999**, *24*, 1215–1217. [CrossRef]
19. Zhang, F.; Zhu, H.T.; Liu, J.; He, Y.F.; Jiang, D.P.; Tang, F.; Su, L.B. Tunable Yb:CaF$_2$-SrF$_2$ laser and femtosecond mode-locked performance based on semiconductor saturable absorber mirrors. *Appl. Opt.* **2016**, *55*, 8359–8362. [CrossRef] [PubMed]
20. Jiang, B.B.; Zheng, L.H.; Jiang, D.P.; Yin, H.D.; Zheng, J.G.; Yang, Q.H.; Cheng, G.F.; Su, L.B. Growth and optical properties of ytterbium and rare earth ions codoped CaF$_2$-SrF$_2$ eutectic solid-solution (RE = Y^{3+}, Gd^{3+}, La^{3+}). *J. Rare Earths* **2021**, *39*, 390–397. [CrossRef]
21. Ma, F.; Su, F.; Zhou, R.; Ou, Y.; Xie, L.; Liu, C.; Jiang, D.; Zhang, Z.; Wu, Q.; Su, L.; et al. The defect aggregation of RE^{3+} (RE = Y, La ~ Lu) in MF$_2$ (M = Ca, Sr, Ba) fluorites. *Mater. Res. Bull.* **2020**, *125*, 110788. [CrossRef]
22. Ruan, F.F.; Yang, L.; Hu, G.; Wang, A.M.; Xue, Y.Y.; Yang, L.L.; Wang, Z.X.; Wu, S.H.; He, Z.L. Luminescence Properties of Dy^{3+} Doped Lanthanum Fluoride Crystal by Multi-crucible Temperature Gradient Technology. *Chin. J. Lumin.* **2021**, *42*, 158–164. [CrossRef]
23. Renaud, E.; Robelin, C.; Heyrman, M.; Chartrand, P. Thermodynamic evaluation and optimization of the (LiF + NaF + KF + MgF$_2$ + CaF$_2$ + SrF$_2$) system. *J. Chem. Thermodyn.* **2009**, *41*, 666–682. [CrossRef]
24. Klimm, D.; Rabe, M.; Bertram, R.; Uecker, R.; Parthier, L. Phase diagram analysis and crystal growth of solid solutions Ca$_{1-x}$Sr$_x$F$_2$. *J. Cryst. Growth* **2008**, *310*, 152–155. [CrossRef]
25. Karimov, D.N.; Komar'kova, O.N.; Sorokin, N.I.; Bezhanov, V.A.; Chernov, S.P.; Popov, P.A.; Sobolev, B.P. Growth of congruently melting Ca$_{0.59}$Sr$_{0.41}$F$_2$ crystals and study of their properties. *Crystallogr. Rep.* **2010**, *55*, 518–524. [CrossRef]
26. Xu, J.W.; Zhou, Y.Z.; Zhou, G.Q.; Xu, K.; Deng, P.Z.; Xu, J. Growth of large-sized sapphire boules by temperature gradient technique (TGT). *J. Cryst. Growth* **1998**, *193*, 123–126.
27. Johnson, L.F.; Guggenheim, H.J. Laser emission at 3 μm from Dy^{3+} in BaY$_2$F$_8$. *Appl. Phys. Lett.* **1973**, *23*, 96–98. [CrossRef]

Article

Study on the Internal Mechanism of APD Photocurrent Characteristics Caused by the ms Pulsed Infrared Laser Irradiation

Liang Chen, Di Wang, Guang-Yong Jin * and Zhi Wei

Jilin Key Laboratory of Solid-State Laser Technology and Application, School of Science, Changchun University of Science and Technology, Changchun 130022, China; 2016200003@mails.cust.edu.cn (L.C.); wangdi@cust.edu.cn (D.W.); weizhi@cust.edu.cn (Z.W.)
* Correspondence: jgyciom@cust.edu.cn

Abstract: In this paper, the sampling current characteristics of the external circuit and the internal mechanism of the current generation in APD irradiated by a millisecond pulse laser were studied. The photocurrent of APD irradiated by a millisecond pulse laser with different energy densities was obtained by the sampling resistance of the external circuit. The photocurrent can be divided into a photocurrent stage, conduction stage and recovery stage in the time domain. This is mainly due to the carrier flow in APD, which leads to the lowering of the barrier between the PN junction. The research results of this paper can be extended to the response of the detector to the high-power infrared pulse laser and provide a certain experimental basis for the design of a millisecond pulse infrared laser detection circuit.

Keywords: photocurrent; carrier flow; lowering of the barrier; APD

Citation: Chen, L.; Wang, D.; Jin, G.-Y.; Wei, Z. Study on the Internal Mechanism of APD Photocurrent Characteristics Caused by the ms Pulsed Infrared Laser Irradiation. *Crystals* **2021**, *11*, 884. https://doi.org/10.3390/cryst11080884

Academic Editors: Xiaoming Duan, Renqin Dou, Linjun Li, Xiaotao Yang and Anna Paola Caricato

Received: 29 June 2021
Accepted: 28 July 2021
Published: 29 July 2021

Publisher's Note: MDPI stays neutral with regard to jurisdictional claims in published maps and institutional affiliations.

Copyright: © 2021 by the authors. Licensee MDPI, Basel, Switzerland. This article is an open access article distributed under the terms and conditions of the Creative Commons Attribution (CC BY) license (https://creativecommons.org/licenses/by/4.0/).

1. Introduction

As the main input port of detection and measurement systems, photodetectors have been widely used in optical tomography, communication, radar and other technical fields [1,2]. Compared to other detectors, an avalanche photodiode (APD) has the advantages of a high quantum efficiency, so this photodetector has been used most widely [3–6]. The multi-pulse 1064-nm laser that irradiated APD was researched, and the results showed that a multi-pulse did not affect the damage threshold when the repetition frequency was less than 1 Hz; however, when the repetition frequency was greater than 1 Hz, the damage threshold decreased with an increase of the number of pulses [7]. In 2015, Koronnov A. A. et al. studied the characteristics of germanium avalanche photodiodes irradiated by a high-powered laser using a 4-ns pulse width and 1.064-um wavelength pulse laser [8]. The experimental and theoretical investigations of millisecond pulse laser ablation-biased Si avalanche photodiodes were studied, and the mechanisms of the phenomenon were studied experimentally and theoretically [9]. The C–V curve of Si-APD was obtained by using a semiconductor analyzer when the millisecond pulse laser was irradiating [10]. This research mainly focused on the research of laser damage to APD, and the law of the change of the photocurrent and the carrier transport behavior in APD and the potential barrier between the PN junctions have not been studied. When APD is used as a laser detection element, it converts the photogenerated current into a voltage signal through a load resistor. Therefore, it is of great significance to study the characteristics of the photocurrent and the mechanism of photocurrent generation when a pulsed infrared laser irradiates APD.

In this paper, the internal mechanism of photocurrent generation in APD irradiated by a millisecond pulse laser is studied for the first time. The microscopic mechanism of the barrier between the carrier and PN junction in APD is studied by the macroscopic representation of the sampling current. With the rapid development of a mid-infrared laser, the detection technology of infrared lasers has gradually attracted people's attention. The research results of this paper can be extended from the response of the detector to high-powered infrared pulse lasers and provide a certain experimental basis for the design of the millisecond pulse infrared laser detection circuit.

2. Materials and Methods

The experimental arrangement is shown schematically in Figure 1. The wavelength of the millisecond laser is 1064 nm (the laser was self-developed by our research group, and the pulse width is 1–5 ms and is tunable), and the pulse repetition rate could be varied manually from 1 Hz to 10 Hz by changing the firing rate of the flash lamp. An attenuator with an attenuation rate of 10 was located at the exit of the laser; the 2:1 Beam:Split was placed after the attenuator. The TEM00 mode laser beam was focused by a lens with a 20-cm focal length. The beam spot size determined by the slit-scan technique was of the order of 360 μm at the focal spot and was independent of the pulse repetition rate, and the error of the spot size was about 5% [11,12]. The Si-APD was biased with 180 V when it was irradiated by the millisecond pulse laser.

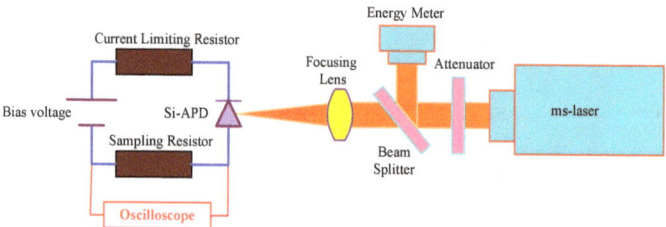

Figure 1. Schematic of the experimental configuration.

The test samples were Si-APDs with the reach-through structure shown in Figure 2. The surfaces of these detectors are coated with a standard antireflection film, and the active area is 0.5 mm^2.

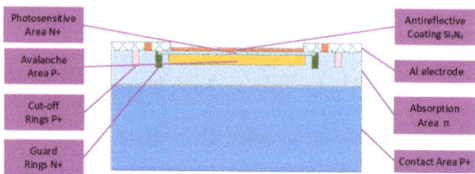

Figure 2. Cross-section of the Si-APD with a reach-through structure.

The experimental system consists of a sampling resistor, current limiting resistor, Si-APD, high-voltage DC power supply, oscilloscope, etc., of which the sampling resistance is 50 KΩ, the current limiting resistor is 1 MΩ and the working voltage is 180 V. The sample is a drawing-type silicon-based APD detector, which is mainly composed of four layers of N$^+$–P$^-$–π–P$^+$. When a higher reverse bias voltage is applied to the outside, the depletion region passes through and reaches the P$^+$ region, and the reverse bias voltage, which exceeds the multiplication voltage, falls all in the π-region; moreover, the collision ionization occurs when the carrier is in the high field region, and they can obtain a sufficiently high average velocity. Since the π-region is much wider than the P-region, the field intensity in the high field region (P-region) and the multiplication rate of the carrier, which are above the avalanche voltage, increase slowly with the reverse bias. Under the working conditions, although the electric field in the π-region is much weaker than in the high field region (P-region), it is sufficient to keep the carrier at a certain drift velocity and transit the π-region in a short time. The faster the carrier drifts, the shorter the transit time is, and the more avalanche ionization occurs when entering the P-region.

3. Results and Discussion

Figure 3 shows that the sampling current with time is divided into three stages: the photogenerated current stage, conduction stage and recovery stage. During the first stage, the peaks of the photocurrent are decreased while the laser energy density increases, and the peak time is basically the same. However, in the conduction stage, the photocurrent is a constant under the action of a laser with different energy densities, and its value depends on the external resistance. In the recovery phase, the recovery time increases with the increase of the laser energy density.

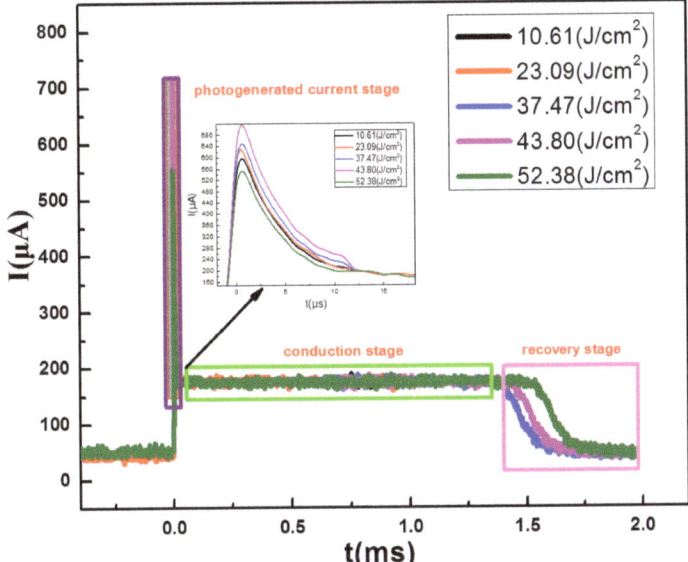

Figure 3. The time-varying diagram of a 1-ms pulse width and 180-V bias voltage at the different energy densities between the sampling current.

The reason for this phenomenon can be explained as follows: with the increasing of the laser energy density, the number of photogenerated electron–hole pairs increased in the π-region, owing to the action of the π-region electric field, the photogenerated carriers, which resulted from avalanche ionization that swept fast into the P-region (multiplication region), and then the inverse current, which is the photocurrent generated in the Si-APD. Unfortunately, due to the external resistance limit, not all of the photogenerated carriers can be quickly exported into the external circuit, and the excess carriers are accumulated, which results in the cumulative effect of a charge at both ends of the PN junction. The accumulated charge forms a photomotive electromotive force at both ends of the PN junction, thereby pulling down the built-in electric field of the π and P regions. Thus, the increase of the reverse current is hindered. With the increase of the incident laser light intensity, the number of electron hole pairs increases at the same time, and the photogenerated electromotive force increases in order to further pull down the barrier in the junction area [13,14]. The schematic diagram of the barrier change is shown in Figure 4. That is to say, the decrease of the potential barrier between the PN junctions in APD is the basic reason that the peak value of the photocurrent decreases with the increase of the laser intensity.

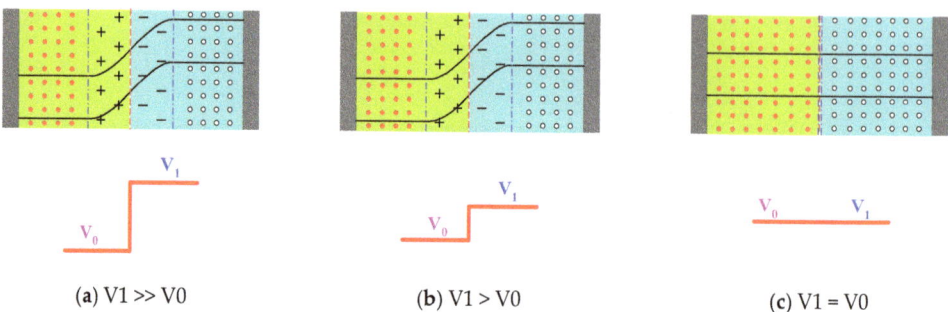

(a) V1 >> V0   (b) V1 > V0   (c) V1 = V0

Figure 4. Schematic diagram of the PN junction barrier in APD changing when the light intensity increases.

As the potential barrier between the PN junctions is further lowered, the external electric field at both ends of APD is further reduced until it is zero and APD is in a reversed-conduction state. At this time, the current flowing through APD is a constant, which is equal to the ratio of voltage and resistance in the external circuit. With the continuous action of the laser, the temperature of APD increases gradually. When the temperature is higher than 500 K, APD will go into a failure state [15]. At this time, the current flowing through APD is still equal to the ratio of voltage and resistance in the external circuit.

When the laser action is over, APD enters the stage of heat recovery, which is dominated by heat dissipation. With the increase of laser energy density, the temperature in APD needs more time to recover to the initial temperature. During this period, the built-in electric field gradually returns to the original state, and the carriers in APD still form a photocurrent due to the external electric field, which leads to a prolongation of the time when the photocurrent drops to zero. The internal mechanism of generating a photocurrent when a millisecond pulse laser irradiates APD can be explained by the following flowchart, as shown in Figure 5 [16].

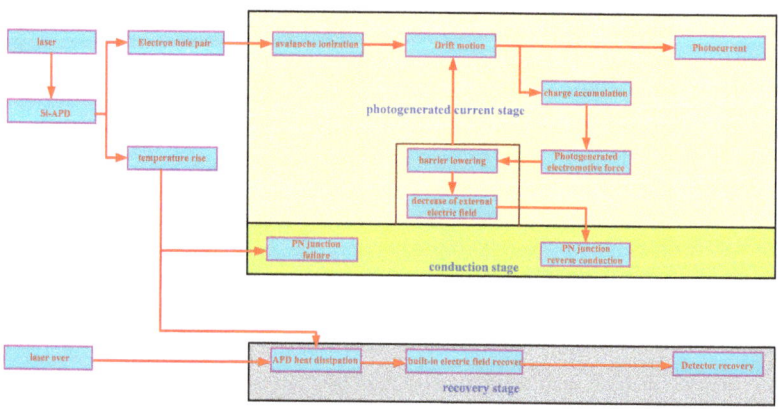

Figure 5. The internal mechanism of generating a photocurrent when the millisecond pulse laser irradiates APD.

4. Conclusions

In this paper, the sampling current characteristics of APD irradiated by a millisecond pulse infrared laser were studied, and the internal generation mechanism of the current characteristics was analyzed. It was found that the sampling current can be divided into a photocurrent stage, conduction stage and recovery stage. In the photocurrent generation stage, the peak current decreases with the increase of the laser energy density, which is mainly due to the decrease of the potential barrier between the PN junctions caused by the accumulation of charges at both ends of the PN junctions. When the barrier between PN junctions dropped to zero, the APD turned on in reverse, and the current value is a certain value, which is equal to the ratio of voltage-to-resistance in the external circuit. In the recovery phase, built-in electric field recovery and temperature recovery are carried out simultaneously; thus, the higher the energy density of the incident laser, the longer it takes for the current to fall to zero.

Author Contributions: Conceptualization, L.C. and Z.W.; methodology, Z.W.; software, D.W.; validation, L.C. and G.-Y.J.; formal analysis, Z.W.; investigation, L.C.; resources, L.C.; data curation, D.W.; writing—original draft preparation, L.C.; writing—review and editing, L.C.; visualization, Z.W.; supervision, G.-Y.J. and project administration and funding acquisition, G.-Y.J. All authors have read and agreed to the published version of the manuscript.

Funding: This study received funding from the National Natural Fund Project of China (grant no. 61805024).

Data Availability Statement: The data presented in this study are available upon request from the corresponding author.

Acknowledgments: We thank the Key Laboratory of Jilin Province Solid-State Laser Technology and Application for use of the equipment.

Conflicts of Interest: The authors declare no conflict of interest.

References

1. Enne, R.; Steindl, B.; Zimmermann, H. Speed optimized linear-mode high-voltage CMOS avalanche photodiodes with high responsivity. *Opt. Lett.* **2015**, *40*, 4400–4403. [CrossRef] [PubMed]
2. Gaberl, W.; Steindl, B.; Schneider-Hornstein, K.; Enne, R.; Reinhard, H.Z. 0.35 lm CMOS avalanche photodiode with high responsivity and responsivity–bandwidth product. *Opt. Lett.* **2014**, *39*, 586–589. [CrossRef] [PubMed]
3. Ferraro, M.S.; Clark, W.R.; Rabinovich, W.S.; Mahon, R.; Murphy, J.L.; Goetz, P.G.; Thomas, L.M.; Burris, H.R.; Moore, C.I.; Waters, W.D.; et al. InAlAs/InGaAs avalanche photodiode arrays for free space optical communication. *Appl. Opt.* **2015**, *54*, F182–F188. [CrossRef] [PubMed]
4. Mu, Y.; Niedre, M. Fast single photon avalanche photodiode-based time-resolved diffuse optical tomography scanner. *Biomed. Opt. Express* **2015**, *6*, 3596–3609. [CrossRef] [PubMed]
5. Nada, M.; Kanazawa, S.; Yamazaki, H.; Nakanishi, Y.; Kobayashi, W.; Doi, Y.; Ohyama, T.; Ohno, T.; Takahata, K.; Hashimoto, T.; et al. High-linearity Avalanche Photodiode for 40-km Transmission with 28-Gbaud PAM4. In *Optical Fiber Communication Conference*; Optical Society of America: Washington, DC, USA, 2015. [CrossRef]
6. Nada, M.; Yokoyama, H.; Muramoto, Y.; Ishibashi, T.; Matsuzaki, H. A 50-Gbit/s vertical illumination avalanche photodiode for 400-Gbit/s Ethernet systems. *Opt. Express* **2014**, *22*, 14681–14687. [CrossRef] [PubMed]
7. Arora, V.K.; Dawar, A.L. Laser-induced damage studies in silicon and silicon-based photo detectors. *Appl. Opt.* **1996**, *35*, 7061–7065. [CrossRef]
8. Koronnov, A.A.; Zverev, G.M.; Zemlyanov, M.M. Characteristics of the germanium avalanche photodiode subjected to a high power laser irradiation. *Prikl. Fiz.* **2015**, *4*, 54–58.
9. Wang, D.; Wei, Z.; Jin, G.-Y.; Chen, L.; Liu, H.-X. Experimental and theoretical investigation of millisecond-pulse laser ablation biased Si avalanche photodiodes. *Int. J. Heat Mass Transf.* **2018**, *122*, 391–394. [CrossRef]
10. Yuan, D.; Di, W.; Zhi, W.; Ran, F.T. Study on the inversion of doped concentration induced by millisecond pulsed laser irradiation silicon-based avalanche photodiode. *Appl. Opt.* **2018**, *57*, 1051–1055. [CrossRef] [PubMed]
11. Anoop, G.; Milster, T.D. Spot distribution measurement using a scanning nanoslit. *Appl. Opt.* **2011**, *50*, 4746–4754.
12. Fair, R.B. A wide slit scanning method for measuring electron and ion beam profiles. *J. Phys. E Sci. Instrum.* **1971**, *4*, 35–36. [CrossRef]
13. Brennan, K.F. *The Physics of Semiconductors*; Cambridge University Press: Cambridge, MA, USA, 1999.
14. Selberherr, S. *Analysis and Simulation of Semiconductor Devices*; Springer: New York, NY, USA, 1984.
15. En-Ke, L.; Bing-Sheng, Z.; Jin-Sheng, L. *The Physics of Semiconductors*; Electronics Industry: Beijing, China, 2011.
16. Neamen, D.A. *Semiconductor Physics and Devices: Basic Principles*; The McMGraw-Hill Education: New York, NY, USA, 2012.

Article

Temperature Rise Characteristics of Silicon Avalanche Photodiodes in Different External Capacitance Circuits Irradiated by Infrared Millisecond Pulse Laser

Liang Chen, Zhi Wei, Di Wang, Hong-Xu Liu and Guang-Yong Jin *

Jilin Key Laboratory of Solid-State Laser Technology and Application, School of Science, Changchun University of Science and Technology, Changchun 130022, China; 2016200003@mails.cust.edu.cn (L.C.); weizhi@cust.edu.cn (Z.W.); wangdi@cust.edu.cn (D.W.); Liuhongxu@mails.cust.edu.cn (H.-X.L.)
* Correspondence: jgyciom@cust.edu.cn; Tel.: +86-431-85582465

Abstract: We experimentally studied the interaction between a millisecond pulse laser and silicon avalanche photodiode (Si-APD) in an external capacitance circuit. The temperature rise law of Si-APD irradiated by a millisecond pulse laser under different external capacitance conditions was obtained. The results show that the surface temperature rise in a Si-APD is strongly dependent on the external capacitance. That is, the smaller the external capacitance, the smaller the surface temperature rise. The effect of the external capacitance on the surface temperature rise in a Si-APD was investigated for the first time in the field of laser damage. The research results have a certain practical significance for the damage and protection of mid-infrared detectors.

Keywords: millisecond pulse laser; silicon avalanche photodiode (Si-APD); external capacitance

Citation: Chen, L.; Wei, Z.; Wang, D.; Liu, H.-X.; Jin, G.-Y. Temperature Rise Characteristics of Silicon Avalanche Photodiodes in Different External Capacitance Circuits Irradiated by Infrared Millisecond Pulse Laser. *Crystals* **2021**, *11*, 866. https://doi.org/10.3390/cryst11080866

Academic Editor: Andrew V. Martin

Received: 25 June 2021
Accepted: 23 July 2021
Published: 26 July 2021

Publisher's Note: MDPI stays neutral with regard to jurisdictional claims in published maps and institutional affiliations.

Copyright: © 2021 by the authors. Licensee MDPI, Basel, Switzerland. This article is an open access article distributed under the terms and conditions of the Creative Commons Attribution (CC BY) license (https://creativecommons.org/licenses/by/4.0/).

1. Introduction

Infrared photodetectors, as an indispensable bridge for the transformation of optical signals into electrical signals, have been widely used in various technical fields [1–4]. In practical applications, the detector is often placed on the focal of the optical receiving system to achieve efficient absorption of light energy. However, after the photodetector absorbs light energy, its temperature rises, greatly affecting its performance. When the detector surface temperature increases, the dark current of the device will increase, thus the signal noise ratio will decrease, and the detection effect of the pairing signal will be affected. When the surface temperature of the detector increases further, it may cause soft damage or hard damage to the detector. Therefore, it is necessary to research the thermal characteristics of the detector. Kruer and Bartoli established a thermal model for polycrystalline lead sulfide (PbS), polycrystalline lead selenide (PbSe), and mercury cadmium telluride (HgCdTe) infrared (IR) detectors [5,6]. Among numerous kinds of detectors, APDs have the unique advantages of small size, high sensitivity, and the ability to work at room temperature. They are thus employed in various fields [7–10], especially in the field of laser detection, where they are most widely used. However, to date, there have been few studies on the new physical phenomena when an APD is irradiated by a millisecond pulse laser. In 2015, with the use of a 1064 nm pulse laser, the characteristics of an irradiated avalanche photodiode were studied [11]. Our research group has studied the temperature rise characteristics of a millisecond pulse laser interacting with a Si-APD and the capacitance–voltage curve of the damaged Si-APD under the condition of no external capacitance. The research results show that Joule heating plays a significant role when the Si-APD is irradiated by a ms pulse laser [12]. In this study, based on previous research, an experimental study on the interaction between the millisecond pulse laser and the Si-APD in an external capacitor circuit was carried out. It was firstly found that the maximum surface temperature of the Si-APD decreases significantly with a decrease in external capacitance. An explanation is given from the point of view of carrier transport.

The existence of external capacitance has a certain blocking effect on the carrier inside the detector. With the capacitor charging process, the carrier transportability becomes weaker and weaker, and the current density inside the PN junction decreases in an exponential form. When the capacitor is fully charged, the circuit is open, and carrier transport process stops. Moreover, the smaller the capacitance value of the external capacitor, the easier it is to fill with charge, and the stronger the hindrance to the transport capacity of the carrier. The results of this research can be used to guide the design of infrared laser detection circuits. The research results have a certain practical significance for the damage and protection of mid-infrared detectors [13,14]. By optimizing the value of capacitance in the driving circuit, the anti-damage ability of the infrared detector towards mid-infrared lasers generated by a periodically polarized crystal can be further improved. When the detector surface temperature increases, the dark current of the device will increase, and thus the signal to noise ratio will decrease, and the detection effect of the pairing signal will be affected. When the surface temperature of the detector increases further, it may cause soft damage or hard damage to the detector. Therefore, it is very significant to study the surface temperature rise in the detector under the action of laser irradiation.

2. Materials and Methods

Figure 1 is the experimental device for the measurement of the temperature rise in a Si-APD irradiated by a millisecond pulse laser. The device is mainly composed of a laser system, a point thermometer, and a Si-APD external circuit. A Nd:YAG millisecond pulse laser with a pulse width was 1.0 ms and a wavelength of 1064 nm is used. The laser irradiates the surface of the detector vertically, and the focus is located on the surface of the detector. The point temperature meter and the optical axis of the laser are placed at 45 degrees, which can effectively avoid the influence of the laser on the measurement accuracy of the point temperature meter. The emitted laser beam is divided into two parts by a beam splitter. One part of the laser beams is focused on the APD surface through a lens, the energy of the other part is monitored by an energy meter, and the ratio of laser energy incident on the surface of the Si-APD detector and energy meter is 10:1. The diameter of the focusing spot is 360 µm. The laser and the point temperature meter are triggered by the same trigger, which can ensure that the point temperature meter begins to measure the temperature at the moment when the laser starts to irradiate, to ensure the synchronicity of the laser-irradiated Si-APD and the point temperature meter. In the experiment, we used a laser frequency of 1 Hz. It is simply a one shot–one temperature measure. The Si-APD external circuit is composed of a 180 V-bias voltage source, capacitor, resistor, and Si-APD. In the experiment, 1 µf and 47 pf capacitors were connected in series in the bias circuit. The Si-APD (the commercial specification is Ø800, made in China) used in the experiment was the $N + P - \pi P^+$ type, with a layered structure, which is composed of four layers, and the doping distribution and thickness of each layer are different. The doping concentrations of each layer are 5×10^{19} cm^{-3}, 5×10^{16} cm^{-3}, 5×10^{12} cm^{-3}, and 1×10^{19} cm^{-3}. The corresponding thicknesses are 1 µm, 3 µm, 50 µm, and 250 µm.

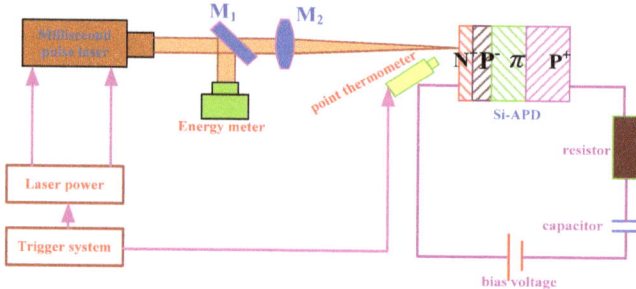

Figure 1. Experimental device.

3. Results

The classical heat conduction process can be expressed as follows [15,16]:

$$k\frac{\partial^2 T(r,z,T,t)}{\partial z^2} + k\frac{\partial^2 T(r,z,T,t)}{\partial r^2} + Q(r,z,T,t) = \rho C\frac{T(r,z,T,t)}{\partial T} \quad (1)$$

where ρ, C, and k are the material density, specific heat, and heat conductivity, respectively. When the Si-APD is placed in the reverse bias circuit and irradiated by the millisecond pulse laser, the heat source term is improved as the sum of the laser heat source outside the detector and the Joule heat source inside the detector, and the equation can be expressed as:

$$Q(r,z,T,t) = Q_L(r,z,T,t) + Q_E(r,z,T,t) \quad (2)$$

Finally, the following equation can be derived:

$$k\frac{\partial^2 T(r,z,T,t)}{\partial z^2} + k\frac{\partial^2 T(r,z,T,t)}{\partial r^2} + Q_L(r,z,T,t) + Q_E(r,z,T,t) = \rho C\frac{T(r,zT,t)}{\partial T} \quad (3)$$

The expression of laser heat is:

$$Q_L(r,z,T,t) = \frac{[1-R(T)] \cdot \alpha(T) \cdot E \cdot g(t)}{\pi r_{las}^2 \tau} \cdot 2\exp(\frac{-2r^2}{r_{las}^2})\exp[-\alpha(T)z] \quad (4)$$

E and τ are the pulse energy and pulse width of the laser, respectively. $\alpha(T)$, $R(T)$, $g(t)$ are the absorption coefficients of silicon, the reflection coefficient of silicon, and the time distribution of the laser beam, respectively. The expression of Joule heat is:

$$Q_E(r,z,T,t) = E(z)J_L(r,z,T,t) \quad (5)$$

When the pulse width is 1.0 ms, energy density is 15 J/cm^2, and theoretical and experimental temperature rise characteristics of the Si-APD in different capacitor driving circuits are shown as Figure 2. Additionally, the system parameters used in the theoretical simulation are as follows: $\rho = 2330 - 2.19 \times 10^{-2} T$ (kg·m^{-3}), specific heat $C = 352.43 + 1.78 T - 2.21 \times 10^{-2} T^2 + 1.3 \times 10^{-6} T^3 - 2.83 \times 10^{-10} T^4$ (J·kg·k^{-1}), $k = 22.23 + 422.52 \times \exp(-T/255.45)$ (W·m^{-1}·k^{-1}).

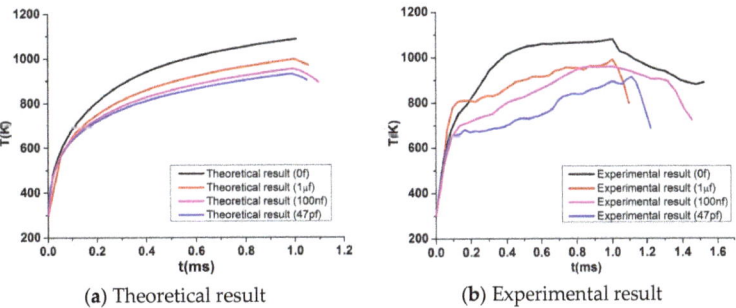

(a) Theoretical result (b) Experimental result

Figure 2. The theoretical and experimental temperature rise characteristics of Si-APD (pulse width 1.0 ms and energy density 15 J/cm^2).

The results show that the temperature change in the Si-APD surfaces can be divided into rapid rise stage, plateau stage, and temperature drop stage. This is because, at the beginning of the laser action, the Si-APD is heated not only by the incident laser, but also by the Joule heat, that is to say, the dual heat sources are working together.

The temperature rise in the Si-APD increases the Joule heat, and the increase in the Joule heat further intensifies the temperature rise. In this way, the surface temperature and internal Joule heat of the Si-APD are mutually iterative, and the surface temperature of the Si-APD increases rapidly in a short time at the beginning of laser irradiation. However, according to the theory of semiconductor physics, when the temperature of the PN junction is more than 520 K, the characteristics of the semiconductor will fail, and the Joule heat in the PN junction will disappear. Therefore, only a single laser source can heat the Si-APD. Under the action of heat conduction, the surface temperature of the Si-APD rises slowly and enters a plateau stage. At the end of the laser action, all heat sources disappear, and the surface temperature decreases rapidly under the action of heat conduction. At the same time, we can also see that the temperature rise is significantly affected by different external capacitors. With the decrease in capacitance, the surface temperature of the Si-APD decreases. This observation shows that the surface temperature of the Si-APD is strongly dependent on the external capacitance.

Based on the study of the temperature rise process under different external capacitance conditions, the maximum temperature rise in the Si-APD surface under different external capacitance conditions was investigated by changing the laser energy density. The experimental results are shown in Figure 3.

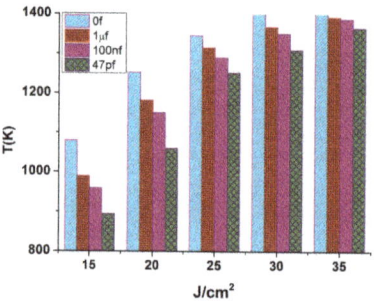

Figure 3. The maximum temperature rise on Si-APD surface (pulse width 1.0 ms and different energy densities).

As shown in the figure, when the laser energy density is constant, the maximum value of the temperature rise without a capacitance is higher than that with a capacitance. When the driving circuit is connected with different capacitors, the maximum temperature rise will be significantly affected. The smaller the capacitance in the driving circuit, the lower the maximum temperature rise on the detector surface. That is to say, the maximum temperature rise on the detector surface is inversely proportional to the capacitance value. To better understand this, we studied the mechanism by analyzing the carrier transport process. Figure 4 shows the internal mechanism for the influence of the different capacitances on the maximum surface temperature rise in the Si-APD. The case without capacitance is shown in Figure 4a, the barrier between PN junctions in the Si-APD is highest under the action of a reverse bias voltage, and a high electric field intensity area is generated in the depletion layer. When the millisecond pulse laser irradiates the Si-APD, the photocarriers generated in the Si-APD are swept away rapidly under the action of high electric field intensity, which results in a large transient current. As Joule heat is the product of electric field intensity and current density, more Joule heat will be generated in the Si-APD when the electric field strength and current density become large simultaneously. However, when a capacitor is connected in series in the circuit, the charges in the circuit will gradually accumulate on both sides of the capacitor plate, hindering the current in the circuit. In this way, the photogenerated carriers in the Si-APD cannot flow into the external circuit smoothly. The carriers accumulate on both sides of the PN junction, forming an additional potential that is opposite to the original potential. This additional potential and the original potential cancel each other out due to the opposite direction, resulting in a decrease in the internal total electric field. Thus, the current density and electric field intensity decrease simultaneously, which reduces the Joule heat. At the same time, it can be seen from Figure 4b,c that the smaller the external capacitance, the weaker its ability to store charges, thus the stronger its ability to block the current in the circuit. At this time, a larger additional potential is formed on both sides of the PN junction, which makes the current density and electric field inside the Si-APD smaller, resulting in less Joule heat. In other words, the Joule heat inside the Si-APD is positively proportional to the capacitance. Therefore, the maximum surface temperature rise in the Si-APD decreases with the decrease in capacitance in the circuit.

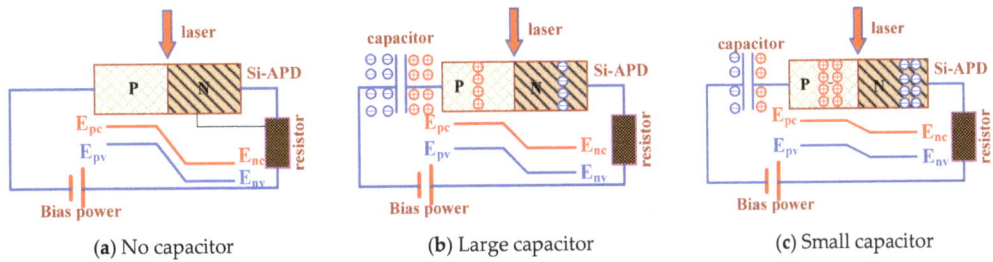

Figure 4. The internal mechanism for the influence of different external capacitances on the maximum temperature rise on Si-APD surfaces.

4. Conclusions

In conclusion, the temperature rise characteristics of the interaction between a millisecond pulse laser and the Si-APD in a capacitor circuit have been studied theoretically and experimentally. The results show that the Joule heat is only produced in a very short time at the beginning of laser action, so the surface temperature change process of a Si-APD under different capacitances can be divided into three stages: rapid heating period, plateau period, and falling period. Under the condition of the same laser parameters and different capacitors, the maximum surface temperature rise in the Si-APD has a significant dependence on the external capacitance, which mainly shows that the maximum temperature rise is smaller with a decrease in capacitance. This result is mainly because the existence of capacitance hinders the generation of photogenerated current in the Si-APD, and with the decrease in capacitance, the blocking effect is more significant. This effect causes the carriers to accumulate on both sides of the PN junction to form an additional potential opposite to the original potential, which reduces the internal field strength of the Si-APD and weakens the Joule heat generation. The results suggest that in the design of a laser detection circuit for APD, the selection of capacitance should not only involve consideration of the filtering effect, but also its influence on the temperature rise characteristics of the Si-APD. In the following experiments, a point temperature meter and multi-spectral temperature meter will be used to measure the temperature jointly, to further improve the accuracy of temperature measurement. This could lay a solid data foundation for further research on the mechanism of the detector surface temperature rise under the action of laser irradiation.

Author Contributions: Coneceptualization: L.C. and D.W.; methodology: L.C.; software: D.W.; validation: L.C., Z.W.; formal analysis: Z.W.; investigation: H.-X.L.; resources: Z.W.; data curation: Z.W.; writing-original draft preparation: L.C.; writing-review and editing: D.W.; visualization: H.-X.L.; supervision: G.-Y.J.; project and funding acquisition: G.-Y.J. All authors have read and agreed to the published version of the manuscript.

Funding: This work was supported by the National Natural Fund Project of China (Grant No. 61805024).

Institutional Review Board Statement: Not applicable.

Informed Consent Statement: Not applicable.

Acknowledgments: We would like to thank the Key Laboratory of Jilin Province Solid-State Laser Technology and Application for permitting us to use their equipment.

Conflicts of Interest: The authors declare no conflict of interest.

References

1. Mccullagh, M.; Wisely, D. 155 Mbit/s optical wireless link using a bootstrapped silicon APD receiver. *Electron. Lett.* **1994**, *30*, 430–432. [CrossRef]
2. Araki, K.; Toyoshima, M.; Takahashi, T.; Fukazawa, T.; Toyoda, M.; Shikatani, M.; Arimoto, Y. Experimental operations of laser communication equipment onboard ETS-VI satellite. In Proceedings of the Free-Space Laser Communication Technologies IX, San Jose, CA, USA, 24 April 1997; pp. 264–276.
3. Krainak, M.A.; Sun, X.; Yang, G.; Lu, W. Comparison of linear-mode avalanche photodiode lidar receivers for use at one-micron wavelength. In *Advanced Photon Counting Techniques IV*; SPIE: Orlando, FL, USA, 2010.
4. Kataoka, J.; Toizumi, T.; Nakamori, T.; Yatsu, Y.; Tsubuku, Y.; Kuramoto, Y.; Enomoto, T.; Usui, R.; Kawai, N.; Ashida, H.; et al. In-orbit performance of avalanche photodiode as radiation detector on board the picosatellite Cute1.7+APD II. *J. Geophys. Res. Atmos.* **2010**, *115*, 1292–1300. [CrossRef]
5. Kruer, M.; Esterowitz, L.; Bartoli, F.; Allen, R. Thermal analysis of laser damage in thin-film photoconductors. *J. Appl. Phys.* **1976**, *47*, 2867–2874. [CrossRef]
6. Bartoli, F.; Esterowitz, L.; Allen, R.; Kruer, M. A generalized thermal model for laser damage in infrared detectors. *J. Appl. Phys.* **1976**, *47*, 2875–2881. [CrossRef]
7. Ferraro, M.S.; Clark, W.R.; Rabinovich, W.S.; Mahon, R.; Murphy, J.L.; Goetz, P.G.; Thomas, L.M.; Burris, H.R.; Moore, C.I.; Waters, W.D.; et al. InAlAs/InGaAs avalanche photodiode arrays for free space optical communication. *Appl. Opt.* **2015**, *54*, F182–F188. [CrossRef] [PubMed]
8. Nada, M.; Kanazawa, S.; Yamazaki, H.; Nakanishi, Y.; Kobayashi, W.; Doi, Y.; Ohyama, T.; Ohno, T.; Takahata, K.; Hashimoto, T.; et al. High-linearity Avalanche Photodiode for 40-km Transmission with 28-Gbaud PAM4. In Proceedings of the Optical Fiber Communication Conference, Los Angeles, CA, USA, 22–26 March 2015.
9. Nada, M.; Yokoyama, H.; Muramoto, Y.; Ishibashi, T.; Matsuzaki, H. 50-Gbit/s vertical illumination avalanche photodiode for 400-Gbit/s Ethernet systems. *Opt. Express* **2014**, *22*, 14681–14687. [CrossRef] [PubMed]
10. Mu, Y.; Niedre, M. Fast single photon avalanche photodiode-based time-resolved diffuse optical tomography scanner. *Biomed. Opt. Express* **2015**, *6*, 3596–3609. [CrossRef] [PubMed]
11. Koronnov, A.A.; Zverev, G.M.; Zemlyanov, M.M.; Zharicova, E.V.; Marsagishvili, D.V. Characteristics of the germanium avalanche photodiode subjected to a high power laser irradiation. *Prikl. Fiz.* **2015**, *4*, 54–58.
12. Wang, D.; Wei, Z.; Jin, G.-Y.; Chen, L.; Liu, H.-X. Experimental and theoretical investigation of millisecond-pulse laser ablation biased Si avalanche photodiodes. *Int. J. Heat Mass Transf.* **2018**, *122*, 391–394. [CrossRef]
13. Peng, Y.; Wang, W.; Wei, X.; Li, D. High-efficiency mid-infrared optical parametric oscillator based on PPMgO: CLN. *Opt. Lett.* **2009**, *34*, 2897–2899. [CrossRef] [PubMed]
14. Dixit, N.; Mahendra, R.; Naraniya, O.P.; Kaul, A.N.; Gupta, A.K. High repetition rate mid-infrared generation with singly resonantoptical parameteric oscillator using multi-grating periodically poled MgO: LiNbO$_3$. *Opt. Laser Technol.* **2010**, *42*, 18–22. [CrossRef]
15. Yang, S.M.; Tao, W.Q. *Heat Transfer*; Higher Education Press: Beijing, China, 2003.
16. Holman, J.P. *Heat Transfer*; The McMGraw-Hill Companies: New York, NY, USA, 2010.

Article

Formation Laws of Direction of Fano Line-Shape in a Ring MIM Plasmonic Waveguide Side-Coupled with a Rectangular Resonator and Nano-Sensing Analysis of Multiple Fano Resonances

Dayong Zhang [1], Li Cheng [2] and Zuochun Shen [1,*]

[1] National Key Laboratory of Tunable Laser Technology, Harbin Institute of Technology, Harbin 150001, China; zhdyhit@163.com
[2] College of Physics and Optoelectronic Engineering, Harbin Engineering University, Harbin 150001, China; chengli@hrbeu.edu.cn
* Correspondence: szc@hit.edu.cn

Abstract: Plasmonic MIM (metal-insulator-metal) waveguides based on Fano resonance have been widely researched. However, the regulation of the direction of the line shape of Fano resonance is rarely mentioned. In order to study the regulation of the direction of the Fano line-shape, a Fano resonant plasmonic system, which consists of a MIM waveguide coupled with a ring resonator and a rectangle resonator, is proposed and investigated numerically via FEM (finite element method). We find the influencing factors and formation laws of the 'direction' of the Fano line-shape, and the optimal condition for the generation of multiple Fano resonances; and the application in refractive index sensing is also well studied. The conclusions can provide a clear theoretical reference for the regulation of the direction of the line shape of Fano resonance and the generation of multi Fano resonances in the designs of plasmonic nanodevices.

Keywords: surface plasmon polaritons (SPPs); metal-insulator-metal (MIM) waveguide; Fano line-shapes; refractive index sensing

1. Introduction

Surface plasmon polaritons (SPPs) are transverse magnetic (TM)-polarized surface waves, which originate from interactions between incident photons and free electrons on metal (like silver and gold) surface and propagate along metal-dielectric interface. SPPs are bound to the metallic surface and decay exponentially in strength with respect to the distance away from the metal-dielectric interface in both directions and can overcome the diffraction limit and control the propagation of the optical signal at the nanoscale [1–3]. In recent times, numerous plasmonic structures have drawn a lot of attention. Among these plasmonic structures, the plasmonic metal-insulator-metal (MIM) waveguide has been studied extensively. It has the advantage of applicable propagation length and good balance between propagation length and losses, and easy integration [4].

In the traditional middle infrared waveguide sensor, the diffraction limit appears when the waveguide size is gradually reduced to the wavelength order, and the wave propagation distance decreases abruptly, which makes greatly reduces the performance of the device. Thus, the traditional middle infrared waveguide sensor cannot achieve miniaturization. Different from the traditional middle infrared waveguide sensor, the plasmonic MIM waveguides can overcome the diffraction limit and can limit the light in the sub-wavelength spatial dimension, as a result, the device is easy to be miniaturized. At the same time, compared with the traditional middle infrared waveguide, the plasmonic MIM waveguide is sensitive to the change in environmental refractive index, so it can be widely used for refractive index sensing. Thus, the plasmonic MIM waveguide has great advantages in refractive index sensing at the subwavelength scale [2,3,5,6].

Fano resonances originate from constructive and destructive interference between localized and continuum states [7,8], which have a significant asymmetric line shape. The wavelength separation between the Fano peak and valley is very small, which offers a highly sensitive spectral response to the variation of the environmental refractive index. Thus, the plasmonic MIM waveguides combined with the Fano resonance effect are widely researched. The research on plasmonic MIM waveguides based on Fano resonance mainly focuses on the creation of Fano resonance by resonators with different geometric shapes [7–14], the influence of the different geometric parameters of these resonators on Fano resonance [7–19], and various analyses of nano-sensing applications [7–19]. The resonators providing continuum mainly include rectangle [9], disk [10], circular ring [6,11], square [12], and various resonators with defects [13]. The resonators providing the localized state mainly include the stub MIM waveguide [6,11,13], single-baffle MIM waveguide [14], and double-baffle MIM waveguide [12,15]. The analyses of nano-sensing applications relate to environmental refractive index [5–14], including temperature sensing [16], chemical and biological sensing [17,18] optical switch [19], and so on. In the above research works, there has been no analysis of the influencing factors of the 'direction' of the Fano line shapes.

In this article, a ring MIM waveguide coupled with a rectangle resonator is proposed to study the influencing factors and formation laws of the 'direction' of the line shape of Fano resonance. It was found that the 'direction' of the Fano line-shape is influenced by different combinations of the orders m of the mode (m, 0) in the ring resonator and n of the mode (n, 0) in the rectangle resonator, and the location of Fano resonance relative to transmission summit of Lorentzian-like line-shape which is dominated by the ring resonator and provides the continuum for Fano resonance. The optimal geometric parameters of the proposed structure for multiple Fano resonances are analyzed and selected to study the refractive index sensing. This study will most likely have important applications for the regulation and control of the Fano line-shape and the generation of multiple Fano resonances in the designs of plasmonic nanodevices.

2. Materials and Methods

Figure 1 shows the proposed plasmonic MIM waveguide system for Fano resonance. The blue and white parts represent silver ($\varepsilon_m = \varepsilon_{Ag}$) and air ($\varepsilon_d = 1.0$), respectively. The width of the MIM waveguide is w. The inner and outer radii of the ring cavity resonator are $R1$ and $R2$ respectively, with the difference between $R1$ and $R2$ fixed at w. The rectangle resonator has a side length of L and a width of w. The coupling distance between the rectangle resonator and the ring resonator is g. The gap between the bus waveguide and the ring resonator is $g1$.

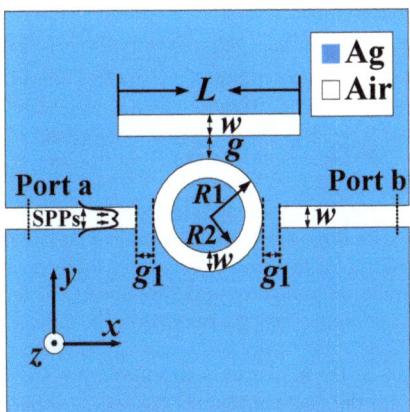

Figure 1. Schematic of the metal-insulator-metal (MIM) plasmonic waveguide system for Fano resonance.

The dielectric constant of Ag is described by the Drude model [20]:

$$\varepsilon_m(w) = \varepsilon_\infty - \frac{w_p^2}{w(w+i\gamma)} \quad (1)$$

where, ε_∞ is the infinite dielectric constant, and the angular frequency ω_p and γ stand is the bulk plasma frequency standing for the natural frequency of the oscillations of free-conduction electrons and the electron collision frequency, respectively. w is the angular frequency of incident light in a vacuum. These parameters for silver can be set as $\varepsilon_\infty = 3.7$, $\omega_p = 1.38 \times 10^{16}$ Hz and $\gamma = 2.73 \times 10^{13}$ Hz.

In the subwavelength MIM plasmonic waveguide, the SSPs propagates in a TM mode. The dispersion equation of the odd mode in the MIM plasmonic waveguide is as follows [2]:

$$\varepsilon_d k_m + \varepsilon_m k_d \tanh\left(\frac{k_d w}{2}\right) = 0 \quad (2)$$

where k_m and k_d are $k_d^2 = \varepsilon_d k_0^2 - \beta^2$ and $k_m^2 = \varepsilon_m k_0^2 - \beta^2$, with ε_d and ε_m defined as dielectric constants of the dielectric medium and the metal, respectively. $k_0 = 2\pi/\lambda_0$ is the wave vector in a vacuum. β represents the complex propagation constant of SPPs. w is the width of the waveguide. The effective refractive index of the plasmonic waveguide is expressed as N_{eff}. The relationship between N_{eff} and β is $N_{eff} = \beta/k_0 = \lambda/\lambda_{spps}$. The N_{eff} can be obtained by solving Equations (1) and (2).

In a rectangular plasmonic cavity, the accumulated phase shift per round trip for the SPPs is $\Phi = 4\pi n_{eff} L_1/\lambda + 2\varphi$. n_{eff} is the real part of N_{eff}. Constructive interference should occur when $\Phi = 2\pi N$, thus the resonance wavelength of mode (0, N) is determined by [21]:

$$\lambda = \frac{2n_{eff} L_1}{N - \varphi/\pi} \quad (3)$$

The physics simulations (transmission spectrums and Hz field distributions) of the proposed structure were investigated by the FEM with COMSOL Multiphysics. The SSPs were excited in the input *port a* and finally collected in the output *port b*. The Fano resonance in our system originates from the interference effect between a localized state caused by the rectangle resonator and a continuum supported by the rectangle resonator. The Fano resonances were studied in the transmittance spectrum. The transmittance, T = $|S_{21}|^2$, is the transmission coefficient from port a to port b.

3. Results and Discussion

3.1. Analysis of the Formation Laws of the 'Direction' of the Fano Line-Shape

In the simulation, these geometric parameters are set as $R1 = 195$ nm, $R2 = 145$ nm, $L = 480$ nm, $w = 50$ nm, and $g = 20$ nm, $g1 = 10$ nm, respectively. To investigate the formation mechanism of Fano resonance, the transmission spectrums of a single ring MIM waveguide, a single rectangle MIM waveguide, and a Fano resonant plasmonic system are shown in Figure 2.

For the single ring MIM waveguide, the transmission spectrums are shown in the dotted blue line, with three transmission peaks of symmetric Lorentzian-like line-shape (providing continuum for the Fano resonator). The corresponding Hz field distributions of the three transmission resonance peaks are given; and the field modes are expressed as ring mode $(m, 0)$ and defined by the number of the dark line in the upper half rings (the Hz field distributions of the upper and lower half ring are symmetric, and only the upper half ring interacts with the rectangle), which are ring modes (2, 0), (3, 0), (4, 0) from right to left, respectively. For the single rectangle MIM waveguide, the transmission spectrums are shown as a dotted red line, with three narrow transmission dips (supporting the located state for the Fano resonator). The corresponding Hz field modes of the three transmission resonance valleys are expressed as rectangle mode $(n, 0)$ and defined by the number of the dark lines in the rectangle, which are rectangle modes (2, 0), (3, 0), (4, 0) from right to left, respectively. The black solid line is the result of the coupling between the continuum and the located state; and the three sharp asymmetric line-shapes in the black solid line are the results of Fano resonance, marked as FR1, FR2, and FR3. The corresponding Hz field distributions of the three Fano resonance peaks exhibit at $\lambda = 376, 485, 713$ nm. From these Hz field distributions, the FR1, FR2, and FR3 originate from the couple between the ring mode (4, 0) and rectangle mode (4, 0), the ring mode (3, 0), and rectangle mode (3, 0), and the ring mode (2, 0) and rectangle mode (2, 0), respectively.

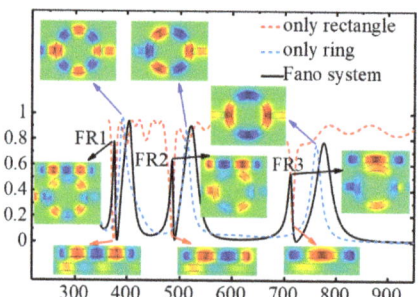

Figure 2. The transmission spectrums of the Fano system with the corresponding Hz field distributions of Fano resonance peaks, of a single ring metal-insulator-metal (MIM) waveguide structure with the corresponding Hz field distributions of transmittance peaks, and of a single rectangle MIM waveguide structure with the corresponding Hz field distributions of transmittance valleys.

In Figure 3, the nephogram ('nephogram' in this paper is not a real nephogram as in meteorology, but is only used to represent a figure of a series of transmission spectrums versus two variables) of Fano resonance phenomenon in the transmission spectrum is exhibited, when the rectangle cavity resonator L changes from 10–1000 nm with the step L is 10 nm and the wavelengths of incident natural light are in the visible and infrared wavelength range.

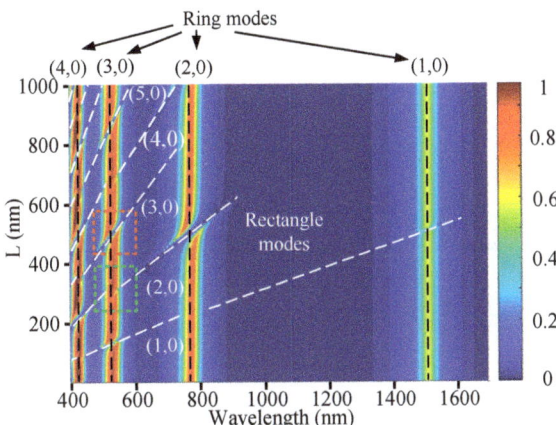

Figure 3. Nephogram of Fano resonance transmission spectrums coupled by metal-insulator-metal (MIM) ring resonator and rectangle resonator with the rectangle resonator L varying from 0 to 1000 nm.

In the nephogram, the black dashed lines represent a series of transmission resonance peaks of different resonance modes in a single ring MIM waveguide. The corresponding resonance modes are ring modes (1,0), (2, 0), (3, 0), (4, 0) from right to left, respectively. The white dashed lines represent a series of transmission resonance peaks of resonance modes in the rectangle MIM waveguide. The corresponding resonance modes are rectangle modes (0,1), (0, 2), ... , and (0, 5), ... , respectively. The red-yellow parts are the transmission spectrums of the interaction between ring modes and ring modes. The split parts of the red-yellow parts (e.g., the parts in dotted red and dotted green boxes) are the results of Fano resonances between the modes in the ring resonator and the rectangle resonator.

Through the analysis, very evident laws are found. When orders m of rectangle modes in the rectangle resonator and n of the ring modes in ring resonator are both odd or both even, the two modes in the ring and rectangle resonators strongly couple with each other, and the corresponding transmission spectrum bands of the Fano resonances are completely split (e.g., the parts of the dotted red box). While, when the orders, m, n, respectively are odd and even (even and odd), the two modes in the ring and rectangle resonators weekly couple with each other and the corresponding transmission spectrum bands of the Fano resonances are incomplete splits (e.g., the part of the dotted green box). Even the Fano resonance coupled by the ring mode (2, 0) and rectangle mode (1, 0) is extremely weak; the corresponding incomplete transmission spectrum band does not get demonstrated.

For further analysis of the 'direction' line shapes of the Fano resonances for the complete and incomplete split transmission spectrum bands, the transmission spectrums in the dotted red and green boxes with different rectangle length L are shown in Figure 4a,b, respectively.

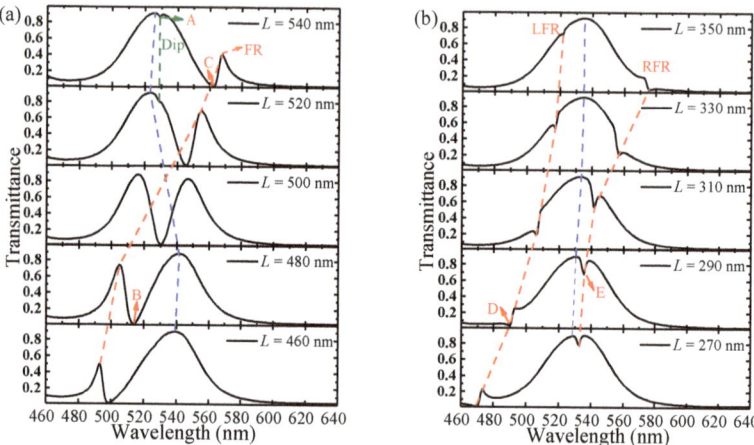

Figure 4. Transmission spectrums for different values of rectangle resonator length L: (**a**) L = 460, 480, 500, 520, and 540 nm, (**b**) L = 270, 290, 310, 330, and 350 nm.

Meanwhile, to deeply understand the physics mechanism of these Fano resonances, Figure 5a–c give the Hz field distributions at the Fano resonance valleys of points 'A'–'E' in Figure 4a,b, respectively.

Firstly, the formula for the line shape of the Fano resonance is analyzed, which is [7,8,22]:

$$\sigma = \frac{(\varepsilon + q)^2}{(\varepsilon^2 + 1)} \qquad (4)$$

q is given by the ratio between the optical response of the located state and the continuum and describes the degree of asymmetry of the resonance. Reduced energy ε is defined by $2(E-E_F)/\Gamma$, with E_F being resonant energy and Γ width of the auto-ionized state. In the limit $|q| \to \infty$, the line-shape is entirely determined by the transition through the discrete state only with the standard 'Lorentzian' profile. The case of $q = 0$ is unique to the Fano resonance and describes a symmetrical dip, sometimes called an anti-resonance. In the case of $q < 0$, the 'direction' of asymmetric Fano line-shape is negative (e.g., the Fano line-shape in Figure 4a with L 460 or 480, where Fano resonance peak is on the left and Fano resonance valley is on the right). In the case of $q > 0$, the 'direction' of asymmetric Fano line-shape is positive (e.g., the Fano line-shape in Figure 4a with L 520 or 540 nm, where Fano resonance peak is on the left and Fano resonance valley is on the right).

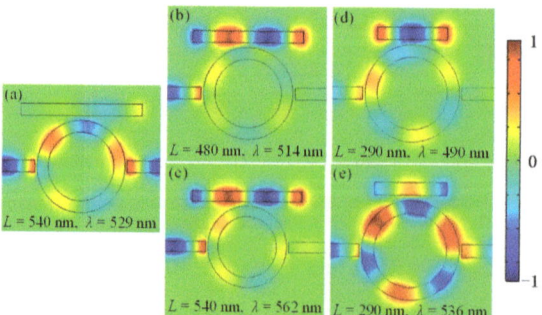

Figure 5. Hz Field distributions of the four points: 'A', 'B', and 'C' (Fano peak) in Figure 3a, 'D' and 'E', (Fano valley) in Figure 3b. (**a**) L = 540 nm, λ = 529 nm, (**b**) L = 480 nm, λ = 514 nm (**c**) L = 540 nm, λ = 562 nm, (**d**) L = 290 nm, λ = 490 nm, and (**e**) L = 290 nm, λ = 536 nm.

In Figure 4a, there are two very weak dips (marked as Dip connected by the dotted green line) and a set of strong (marked as FR connected by the dotted red line) Fano resonances. The very weak dips only occur with L = 520 and 540 nm on the right side of the transmission summit of symmetric Lorentzian-like line-shape. The symmetric Lorentzian-like line-shape is dominated by the ring resonator and provides the continuum for the Fano resonance, the corresponding transmission summits connected by the dotted blue line in Figure 4a. The two very weak dips originate from the couple between the ring odd modes (3, 0) and very weak rectangle even modes (3, 0), which can be known from the Hz field distributions of points 'A' in Figure 5a with L = 540 nm. The strong Fano resonances couple by rectangle odd modes (3, 0) and ring odd modes (3, 0). The corresponding Hz field distributions of points 'B' and 'C' are shown in Figure 5a with L = 480 and 540 nm. It can be seen that the phases of the two rectangular modes (3, 0) are the same, while the two ring modes (3, 0) have some differences. With the increase of L, the Fano resonance shows a red shift and moves from the left side of the transmission summit of symmetric Lorentzian-like line-shape (with the 'direction' of the Fano line-shape $q < 0$) to the right side of the transmission summit (with the 'direction' of the Fano line-shape $q > 0$). In addition, it can be noted that the phase of the weak rectangle even mode (3, 0) of the dip is different from that of the Fano resonance valley in Figure 5a with L = 540 nm. In Figure 4b, these relatively weak Fano resonances couple by rectangle modes (2, 0) and ring modes (3, 0) can be seen. It is to be noted that no matter how big L is, when L is a certain value the rectangle mode (3, 0) simultaneously contributes two Fano resonances (connected by dotted red lines and respectively marked as LFR and RFR). These are respectively located on the left side (with the 'direction' of the Fano line-shape $q > 0$) and right side (with the 'direction' of the Fano line-shape $q < 0$) of the same transmission summit (connected by a dotted blue line) of symmetric Lorentzian-like line-shape. With the L increasing, the two kinds of Fano resonances either show a red shift (the two sets of Fano resonances are marked as LFR and RFR). The corresponding Hz field distributions of points 'D' and 'E' in Figure 4b are both shown in Figure 5b with L = 290 nm. It can be seen that the phases of the two ring modes (3, 0) are the same, while the two rectangular modes are both modes (3, 0) with a phase difference of π. The results are very interesting. According to Equation (3), in the fixed-size rectangle cavities, although the phase is different the same mode should have the same peak position. However, our results demonstrate in the same size rectangle cavities, when the rectangle cavity couples with the ring MIM waveguide shown in Figure 1, the phases of modes can influence the Fano peak positions. The conclusion is suitable for other couples between the ring odd and rectangle even (rectangle even and ring odd) modes, as shown in Figure 3. Unfortunately, the intensities of these Fano resonances in Figure 4b are obviously weaker than those in Figure 4a.

Combined with the analyses of Figures 3–5, we can summarize and generalize the general law of formation of the 'direction' of the Fano line-shape. When the orders m of the mode (m, 0) in ring resonator and n of the mode (n, 0) in rectangle resonator are both odd or both even, and the Fano resonance occurs on the location of the increasing slope of the symmetric Lorentzian-like line-shape with wavelength increasing (the left side of the transmission summit of Lorentzian-like line-shape), the 'direction' of the Fano line-shape is negative, $q < 0$; while and the Fano resonance occurs on the location of decreasing slope of the symmetric Lorentzian-like line-shape with wavelength increasing (the right side of the transmission summit of Lorentzian-like line-shape), the 'direction' of Fano line-shape is positive, $q > 0$. For the condition of the shape parameter $q = 0$ (e.g., Figure 4a with L = 500 nm), the resonance frequencies of the ring and rectangle are equal. When the orders m of the mode (m, 0) in ring resonator and n of the mode (n, 0) in rectangle resonator are respectively odd and even (even and odd), and the Fano resonance occurs on the location of the increasing slope of the symmetric Lorentzian-like line-shape with the wavelength increasing (the left side of the transmission summit) and the 'direction' of the Fano line-shape positive, $q > 0$; while the Fano resonance occurs on the location of decreasing slope of the symmetric Lorentzian-like line-shape with wavelength increasing (the right side of the transmission summit) and the 'direction' of the Fano line-shape negative, $q < 0$.

3.2. Analysis of Optimal Condition for the Generation of Multiple Fano Resonances

Figure 6a shows the nephogram of the transmission spectrum of the proposed Fano system with the wavelength varying from 400 to 1000 nm and the coupling distance g changing from 10 to 80 nm with the other parameters kept unchanged. Figure 6b shows the transmission spectrums from Figure 6a with g = 20, 30, and 40 nm, respectively.

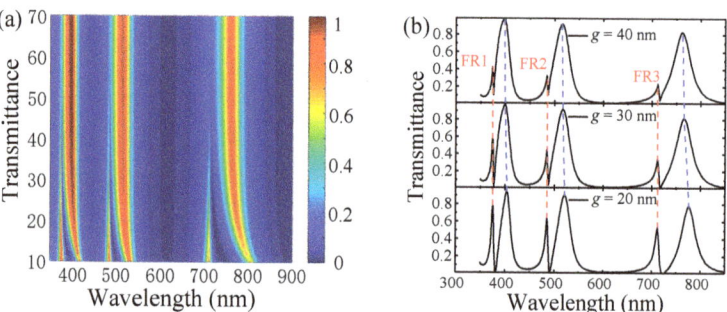

Figure 6. (a) Nephogram of transmission spectrums of the proposed Fano system as a function of g and wavelength with other parameters keep unchanged. (b) Transmission spectrums from Figure 5a with g = 10, 20, and 40 nm, respectively.

As can be seen, when g is bigger (e.g., g = 70 nm), the coupling has not happened yet. As g decreases, the coupling of the two resonators becomes stronger, resulting in modes splitting, and the FR1, FR2, and FR3 (the three narrowly red-yellow spectral bands from left to right) emerge and become stronger. The smaller g is, the more obvious the extent of mode splitting is. With the g decreasing, the resonance wavelength of FR1, FR2, and FR3 is almost unchanged, while the Lorentzian-like line-shapes connected by the dotted blue lines in Figure 6b (the three broadly red-yellow spectral bands in Figure 6a) show an obvious blue shift. Furthermore, combined with the analyses in Figures 3–5, the optimal condition for the generation of multiple Fano resonances is obtained. The optimal condition is that the effective length of the ring resonator is close to that of the rectangle resonator with a reasonable and smaller coupling distance g. At this condition, for the same modes (the orders m and n are the same) the resonance frequency of ring resonance is close to that of rectangle resonance, resulting in a stronger Fano resonance. Specifically, when L = 480 nm the effective length of ring resonance is close to that of a rectangle resonator with g = 20 nm in Figure 2 or Figure 4. Thus, for the modes that the orders m and n are the same ($m = n$ = 2, 3, or 4) the resonance frequency of ring resonance is close to that of rectangle resonance, with three stronger Fano resonances. Compared to a single Fano resonance, a plasmonic spectrum line, which contains multiple Fano resonances, can be simultaneously adjusted at several multiple different spectral locations. And it would be applied to multiwavelength surface-enhanced Raman scattering (SERS), and biosensing [23].

3.3. Nano-Sensing Analysis

In Figure 7a, the impact of refractive index n in the rectangle resonator is also studied with the optimal condition for the generation of multiple Fano resonances (the geometric parameters are consistent with those in Figure 2). As n increases from 1 to 1.02, the wavelength responses can be adjusted from 375.5 nm to 380.2 nm for FR1, from 485.4 nm to 493.4 nm for FR2, and from 712.6 nm to 722.9 nm for FR3, demonstrating a significant sensitivity to the change of environmental refractive index. To better evaluate the performance of refractive index sensing of the Fano resonance, the dimensionless figure of merit (FOM) is studied. The FOM at the frequency λ can be represented as [18]:

$$\text{FOM} = \frac{\Delta T}{T\Delta n} = \frac{T(\lambda, n) - T(\lambda, n_0)}{T(\lambda, n_0)\Delta n} \tag{5}$$

where $T(\lambda, n_0)$ is the original value of transmission, and $T(\lambda, n_0)$ is the transmission after changing the environmental refractive index. $\Delta n = n - n_0$, n_0 is the environmental refractive index in the original state, and n is the environmental refractive index after changing.

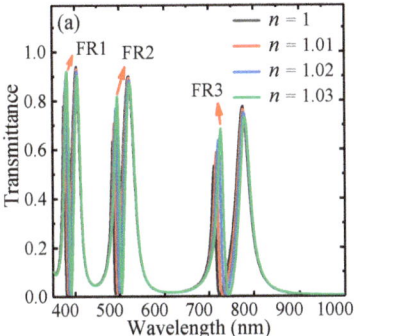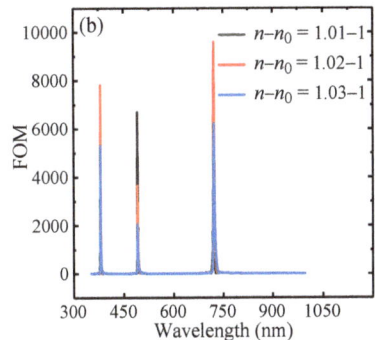

Figure 7. (a) Transmission spectrums with n from 1 to 1.03. (b) Figure of merit (FOM) of FR1, FR2, and FR3 with different n.

Figure 7b shows the computed FOMs with n 1.01, 1.02, and 1.03, where the max FOM can reach 9591 at 750 nm with n = 1.02, which is greater than the max FOM = 3200, 5500, and 6838 found in [6,24,25], respectively.

4. Conclusions

We investigate and find the influencing factors and formation laws of the direction of the Fano line-shape, and the optimal condition for the generation of multiple Fano resonances in a Fano system consisting of a ring MIM waveguide coupled with a rectangle resonator. The rule is as follows. When the orders m of the mode $(m, 0)$ in ring resonator and n of the mode $(n, 0)$ in rectangle resonator are both odd or both even, whether the Fano resonance occurs on the left or right of the transmission summit of a symmetric Lorentzian-like line-shape dominated by ring resonator and providing continuum for Fano resonance, the Fano resonance peak is farther from the transmission summit of the Lorentzian-like line-shape than the Fano resonance valley. When the orders m of the mode $(m, 0)$ in the ring resonator and n of the mode $(n, 0)$ in the rectangle resonator are odd and even, respectively (even and odd respectively), whether the Fano resonance occurs on the left or right of the transmission summit of a symmetric Lorentzian-like line-shape dominated by ring resonator and providing continuum for Fano resonance, the Fano resonance valley is farther from transmission summit of the Lorentzian-like line-shape than the Fano resonance peak. The optimal condition for the generation of multiple Fano resonances is that the effective length of the ring resonator is close to that of the rectangle resonator with a reasonable and smaller coupling distance g. The conclusions can provide a theoretical foundation for the formation laws of the directional of the line shape of Fano resonance and the generation of multiple Fano resonances in the designs of plasmonic nanodevices. Furthermore, because the Fano resonance peaks of the proposed system are very sensitive to the environmental refractive index, when the plasmonic MIM guide and the ring and rectangle resonators are filled by high refractive index material like the study in the paper [26], the Fano resonance phenomenon and the nano-sensing of the environmental refractive index can be extended to the mid-infrared wavelength range.

Author Contributions: Conceptualization and methodology, Z.S., D.Z. and L.C.; software, D.Z. and L.C.; writing—original draft preparation, Z.S., D.Z. and L.C.; writing—review and editing, Z.S. and D.Z. All authors have read and agreed to the published version of the manuscript.

Funding: This research received no external funding.

Data Availability Statement: The data in this study are available from the corresponding authors upon request.

Conflicts of Interest: The authors declare no conflict of interest.

References

1. Barnes, W.L.; Dereux, A.; Ebbesen, T.W. Surface plasmon subwavelength optics. *Nature* **2003**, *424*, 824–830. [CrossRef]
2. Maier, S.A. Plasmonics: Fundamentals and Applications. In *Plasmonics: Fundamentals and Applications*; Springer Science & Business Media: Cham, Switzerland, 2007; pp. 1–9.
3. Ebbesen, T.W.; Lezec, H.J.; Ghaemi, H.F.; Thio, T.; Wolff, P.A. Extraordinary optical transmission through sub-wavelength hole arrays. *Nature* **1998**, *391*, 1114–1117. [CrossRef]
4. Ebbesen, T.W.; Genet, C.; Bozhevolnyi, S.I. Surface-plasmon circuitry. *Phys. Today* **2008**, *61*, 44–50. [CrossRef]
5. Mizaikoff, B. Waveguide-enhanced mid-infrared chem/bio sensors. *Chem. Soc. Rev.* **2013**, *42*, 8683–8699. [CrossRef] [PubMed]
6. Chen, Z.; Yu, L.; Wang, L.; Duan, G.; Zhao, Y.; Xiao, J. Sharp Asymmetric Line Shapes in a Plasmonic Waveguide System and its Application in Nanosensor. *J. Lightwave Technol.* **2015**, *33*, 3250–3253. [CrossRef]
7. Miroshnichenko, A.E.; Flach, S.; Kivshar, Y.S. Fano resonances in nanoscale structures. *Rev. Mod. Phys.* **2010**, *82*, 2257–2298. [CrossRef]
8. Fano, U. Effects of Configuration Interaction on Intensities and Phase Shifts. *Phys. Rev.* **1961**, *124*, 1866–1878. [CrossRef]
9. Wen, K.; Hu, Y.; Chen, L.; Zhou, J.; Lei, L.; Guo, Z. Fano Resonance with Ultra-High Figure of Merits Based on Plasmonic Metal-Insulator-Metal Waveguide. *Plasmonics* **2015**, *10*, 27–32. [CrossRef]
10. Yun, B.F.; Zhang, R.H.; Hu, G.H.; Cui, Y.P. Ultra Sharp Fano Resonances Induced by Coupling between Plasmonic Stub and Circular Cavity Resonators. *Plasmonics* **2016**, *11*, 1157–1162.
11. Zhang, Z.D.; Wang, R.B.; Zhang, Z.Y.; Tang, J.; Zhang, W.D.; Xue, C.Y.; Yan, S.B. Electromagnetically Induced Transparency and Refractive Index Sensing for a Plasmonic Waveguide with a Stub Coupled Ring Resonator. *Plasmonics* **2017**, *12*, 1007–1013. [CrossRef]
12. Chen, Z.; Yu, Y.; Wang, Y.; Guo, N.; Xiao, L. Compact Plasmonic Structure Induced Mode Excitation and Fano Resonance. *Plasmonics* **2020**, *15*, 2177–2183. [CrossRef]
13. Yang, X.; Hua, E.; Su, H.; Guo, J.; Yan, S. A Nanostructure with Defect Based on Fano Resonance for Application on Refractive-Index and Temperature Sensing. *Sensors* **2020**, *20*, 4125. [CrossRef]
14. Su, C.; Zhu, J. Novel SPR Sensor Based on MIM-based Waveguide and an Asymmetric Cross-shaped Resonator. *Plasmonics* **2021**, *16*, 769–775. [CrossRef]
15. Chen, Y.; Luo, P.; Liu, X.; Di, Y.; Han, S.; Cui, X.; He, L. Sensing performance analysis on Fano resonance of metallic double-baffle contained MDM waveguide coupled ring resonator. *Opt. Laser Technol.* **2018**, *101*, 273–278. [CrossRef]
16. Shahamat, Y.; Vahedi, M. Mid-infrared plasmonically induced absorption and transparency in a Si-based structure for temperature sensing and switching applications. *Opt. Commun.* **2019**, *430*, 227–233. [CrossRef]
17. Zhu, J.; Li, N. MIM waveguide structure consisting of a semicircular resonant cavity coupled with a key-shaped resonant cavity. *Opt. Express* **2020**, *28*, 19978–19987. [CrossRef] [PubMed]
18. Becker, J.; Trügler, A.; Jakab, A.; Hohenester, U.; Sönnichsen, C. The Optimal Aspect Ratio of Gold Nanorods for Plasmonic Bio-sensing. *Plasmonics* **2010**, *5*, 161–167. [CrossRef]
19. Wang, Y.L.; Li, S.L.; Zhang, Y.Y.; Yu, L. Ultrasharp Fano Resonances Based on the Circular Cavity Optimized by a Metallic Nanodisk. *IEEE Photonics J.* **2016**, *8*, 8. [CrossRef]
20. Lin, X.-S.; Huang, X.-G. Tooth-shaped plasmonic waveguide filters with nanometeric sizes. *Opt. Lett.* **2008**, *33*, 2874–2876. [CrossRef] [PubMed]
21. Chen, J.; Sun, C.; Gong, Q. Fano resonances in a single defect nanocavity coupled with a plasmonic waveguide. *Opt. Lett.* **2014**, *39*, 52–55. [CrossRef]
22. Gallinet, B.; Martin, O.J.F. Influence of Electromagnetic Interactions on the Line Shape of Plasmonic Fano Resonances. *ACS Nano* **2011**, *5*, 8999–9008. [CrossRef] [PubMed]
23. Liu, N.; Mukherjee, S.; Bao, K.; Li, Y.; Brown, L.V.; Nordlander, P.; Halas, N.J. Manipulating Magnetic Plasmon Propagation in Metallic Nanocluster Networks. *ACS Nano* **2012**, *6*, 5482–5488. [CrossRef] [PubMed]
24. Yi, X.C.; Tian, J.P.; Yang, R.C. Tunable Fano resonance in MDM stub waveguide coupled with a U-shaped cavity. *Eur. Phys. J. D* **2018**, *72*, 9. [CrossRef]
25. Liu, D.D.; Wang, J.C.; Zhang, F.; Pan, Y.W.; Lu, J.; Ni, X.W. Tunable Plasmonic Band-Pass Filter with Dual Side-Coupled Circular Ring Resonators. *Sensors* **2017**, *17*, 585. [CrossRef]

26. Esteban, O.; Gonzalez-Cano, A.; Mizaikoff, B.; Diaz-Herrera, N.; Navarrete, M.C. Generation of Surface Plasmons at Waveguide Surfaces in the Mid-Infrared Region. *Plasmonics* **2012**, *7*, 647–652. [CrossRef]

Article

Experimental Investigation of Double-End Pumped Tm, Ho: GdVO$_4$ Laser at Cryogenic Temperature

Yanqiu Du [1,2], Tongyu Dai [3], Hui Sun [2], Hui Kang [2], Hongyang Xia [2], Jiaqi Tian [2], Xia Chen [2] and Baoquan Yao [3,*]

[1] Key Laboratory of In-Fiber Integrated Optics of Ministry of Education, College of Science, Harbin Engineering University, Harbin 150001, China; hellen_q@126.com
[2] School of Electric and Information Engineering, Heilongjiang University of Science and Technology, Harbin 150022, China; huihuihit@sina.com (H.S.); qihang1601@sina.com (H.K.); xhy04540451@163.com (H.X.); Tjq960616@126.com (J.T.); cx1214658352@126.com (X.C.)
[3] National Key Laboratory of Tunable Laser Technology, Harbin Institute of Technology, Harbin 150001, China; daitongyu2006@126.com
* Correspondence: yaobq@hit.edu.cn

Citation: Du, Y.; Dai, T.; Sun, H.; Kang, H.; Xia, H.; Tian, J.; Chen, X.; Yao, B. Experimental Investigation of Double-End Pumped Tm, Ho: GdVO$_4$ Laser at Cryogenic Temperature. *Crystals* **2021**, *11*, 798. https://doi.org/10.3390/cryst11070798

Academic Editors: Haohai Yu, M. Ajmal Khan, Julien Brault and Andrew V. Martin

Received: 27 May 2021
Accepted: 5 July 2021
Published: 8 July 2021

Publisher's Note: MDPI stays neutral with regard to jurisdictional claims in published maps and institutional affiliations.

Copyright: © 2021 by the authors. Licensee MDPI, Basel, Switzerland. This article is an open access article distributed under the terms and conditions of the Creative Commons Attribution (CC BY) license (https://creativecommons.org/licenses/by/4.0/).

Abstract: We describe comparatively cryogenically cooled Tm, Ho: GdVO$_4$ lasers with an emission wavelength of 2.05 µm under continuous wave and pulse operating mode. By varying the transmittance of output couplers to be 0.40 for a continuous wave laser, the maximum output power of 7.4 W was generated with a slope efficiency of 43.3% when the absorbed pump power was increased to 18.7 W. For passively Q-switched lasers, the output characteristics were researched through altering pump mode radius. When the pump mode radius focused into the Tm, Ho: GdVO$_4$ center equaled near 600 µm, the peak power was increased to be the maximum value of 9.9 kW at the absorbed pump power of 11.8 W. The pulse energy of 0.39 mJ was achieved at the same absorbed pump power with repetition of 5.7 kHz.

Keywords: solid-state; diode-pumped; Q-switched; infrared and far-infrared lasers

1. Introduction

Solid-state pulse lasers at a ~2 µm wavelength band from Tm: $^3F_4 \rightarrow {}^3H_6$ and Ho: $^5I_7 \rightarrow {}^5I_8$ energy levels are attractive due to the extensive application of laser lidar systems to achieve environmental detection and for medical instruments to conduct surgical treatment [1,2]. In addition, the pulse lasers at ~2 µm can also be utilized as pump sources of OPOs to obtain the mid-IR laser [3–5]. Tm/Ho ions doped into different hosts can generate laser emitting at ~2 µm. The vanadate crystal hosts possess large phonon coupling energy, low crystal symmetry, and great thermal conductivity [6]. Especially compared with the YVO$_4$ crystal, the GdVO$_4$ crystal exhibits higher thermal conductivity and lower thermal expansion coefficient that is beneficial for weakening thermal effects [7]. Rare earth ions such as Tm^{3+} and Ho^{3+} doped into vanadate hosts exhibit a broadly spectral band at a pump wavelength of ~800 nm and great absorption/emission cross sections [8] which facilitate the laser crystal to efficiently absorb pump light, thus improving laser performances. Tm-doped vanadate laser [9] pumped by laser diodes with ~800 nm wavelength, Ho-doped vanadate laser [10] in-band pumped by Tm-lasers or laser diodes, and Tm, Ho-codoped vanadate lasers [11,12] pumped by laser diodes have been extensively investigated under continuous wave or pulse operating modes. Tm, Ho-codoped lasers with sensitization ions Tm^{3+} and activation ions Ho^{3+} can be pumped by laser diodes with ~800 nm wavelengths and simultaneously exhibit excellent Q-switching performance due to the strong stored energy capability of Ho^{3+} ions. Besides, the performances of Tm, Ho-codoped vadanate lasers depend strongly on the temperature of active media. At cryogenic temperature, Tm, Ho-codoped lasers operate under the quasi-four level system which is beneficial to reduce upconversion losses ($^5I_7$, $^3F_4 \rightarrow {}^5I_5$, $^3H_6$) of the Tm, Ho-codoped crystals.

Furthermore, the positive energy transfer process between $Tm^{3+}(^3F_4)$ and $Ho^{3+}(^5I_7)$ was significantly strengthened which obviously increases the population proportion reserved in $^5I_7$ energy level.

Passively Q-switched lasers with saturable absorbers usually exhibit some advantages of compact, simple geometry, and low cost. The Cr-doped ZnS crystal is one of the typical SAs with a large optical damage threshold [13] that enables it to undertake strong oscillating laser intensity. The Cr:ZnS crystal also has great absorption/emission cross-sections at ~2 µm [14] which facilitate to get excellent Q-switching performances for the Tm^{3+}/Ho^{3+} doped lasers with an emitting wavelength of ~2 µm [15–17]. Based on Cr^{2+}:ZnS SA, Tm, Ho-codoped laser with different hosts such as LLF [18], YLF [19], and KLu $(WO_4)_2$ [20] have been presented. At cryogenic temperature, we also had demonstrated PQS Tm, Ho-codoped vanadate lasers with Cr:ZnS SA [21,22]. The highest peak power of ~9.1 kW was achieved with the pulse duration of 32.7 ns [23]. In this paper, to improve the peak power, the comparative research for the PQS lasers with different pump mode radius was experimentally conducted.

2. Experimental Setup

The laser (shown in Figure 1) was designed to be a U-shape resonator that makes it possible to pump a laser crystal from each end of the crystal. The pump beams from two fiber-coupled (400 µm diameter, 0.22 numerical aperture) LDs emitting at ~800 nm were collimated and focused by plano-convex lens (F1, F2) into the active crystal from each end of it. The focused beam radius was 400 µm or 600 µm depending to the focus length of focus len (F1). The output mirror (M2) was a plane mirror with different transmittance (T_{oc}) of 0.20, 0.25, and 0.40 at ~2 µm. The mirror (M1) with the 2000 mm curvature was coated in anti-reflective film at ~800 nm and coated in high-reflective film at ~2 µm. The dichroic mirrors (M3, M4) placing at 45° relatively to the oscillating light path were anti-reflective at ~800 nm and high-reflective at ~2 µm. The distances between M2 and M4, M4 and M3, and M3 and M1 were 35 mm, 60 mm, and 45 mm, respectively. The total cavity length is at 140 mm. To improve the absorbed pump power without crystal fracture, an 8 mm long Tm, Ho: $GdVO_4$ crystal (a-cut) with a 4 × 4 mm^2 cross section was chosen as the gain medium with 4at.% Tm^{3+} and 0.4at.% Ho^{3+} concentration. The Tm, Ho: $GdVO_4$ was wrapped in indium foil and held in copper heat-sinks connected with a small dewar. During the operating of the laser, the dewar was full of liquid-N2 to effectively remove the heat generated in the crystal for high power operation. The M5 and M6 in Figure 1 denoted the plane windows with anti-reflective film at ~800 nm and ~2 µm. The distance between M5 and M6 is 22 mm. Under the pump laser beam radius of 400 µm or 600 µm, the actual measured absorption efficiency of the Tm, Ho: $GdVO_4$ crystal was over 90%. A Cr:ZnS SA (2 mm thickness, 9 × 9 mm^2 cross section) with the unsaturated transmission of ~82% was used as Q-switch crystal placed in a brass heat-sink filled with flowing water. It was inserted between the dichroic mirror M3 and the end mirror M1 with 15 mm distance from the mirror M1.

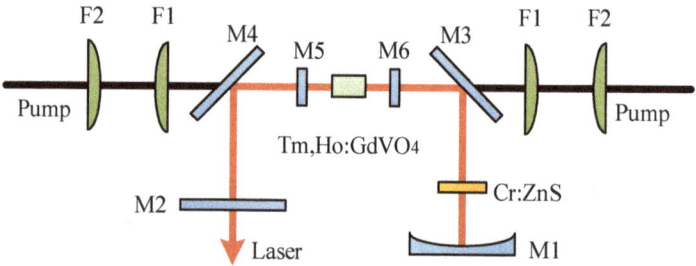

Figure 1. Sketch of experimental setup.

3. Experimental Results and Discussion

At first, focusing the pump laser beam into Tm, Ho: GdVO$_4$ crystal with a near 400 μm mode radius, we conducted experimental research on Tm, Ho: GdVO$_4$ lasers. Through changing output couplers with different transmittance, the CW and PQS lasers were characterized. To avoid the fracture of gain crystal, the total absorbed pump power did not exceed 18.7 W. The output powers for the CW and PQS lasers were shown in Figure 2 with fitted lines. When the transmittance of OC equals 0.40, the maximum output power for CW and PQS operating mode were achieved to be 7.4 W (η = 43.3%) and 5.4 W (η = 32.8%), η denoted the slope efficiency.

Since the high transmittance of OC caused the large resonator loss, the oscillating laser intensity in the resonator was too weak to enable the SA bleached to generate stable Q-switching phenomenon at low pump power. The stable Q-switching for the laser with 0.40 transmittance OC was not achieved until the absorbed pump power was up to 10.7 W. In comparison with this case, the PQS laser with OC transmittance of 0.20 can only generate stable Q-switched pulse trains at the absorbed pump power of less than 10.1 W. The PQS laser with transmittance of 0.25 can achieve a stable pulse operating within the entire range of absorbed pump power (3.5–18.7 W). Furthermore, the laser generated moderate CW output power of 6.2 W (η = 35.6%) and average output power of 4.5 W (η = 26.5%) at the absorbed pump power of 18.7 W.

The minimum pulse duration of ~36 ns was achieved for the PQS lasers with a transmittance of 0.20 and 0.25 when the absorbed pump power was increased to the respective maximum value of 10.1 W and 18.7 W, as shown in Figure 3a. The PQS lasers with transmittance of 0.25 and 0.40 generated approximate pulse energy, 0.26 mJ and 0.28 mJ, when the absorbed pump power was increased to 18.7 W, as shown in Figure 3b. The roughly equivalent repetition rates (shown in Figure 4a) were almost increased linearly with the increase of pump power. In comparison, the peak power of ~7.3 kW was achieved for the PQS laser with transmittance of 0.25 at 18.7 W absorbed pump power (shown in Figure 4b), larger than that of other two cavities. Compared with reference [23] with an approximate geometry, the pulse performance was slightly inferior due to the oscillating mode in the resonator.

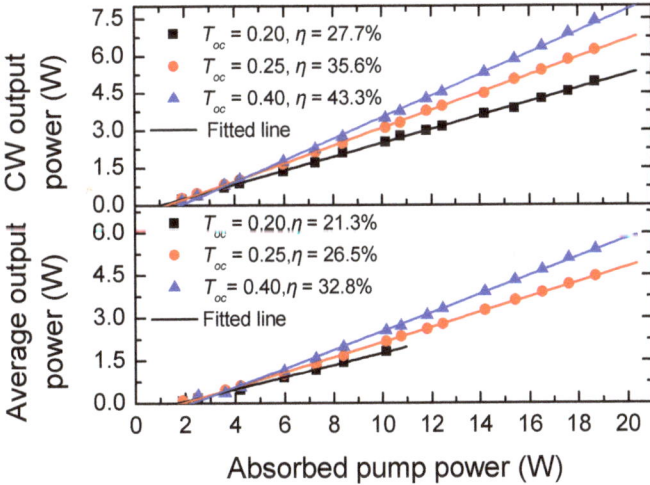

Figure 2. Output power of CW and PQS lasers vs. absorbed pump power.

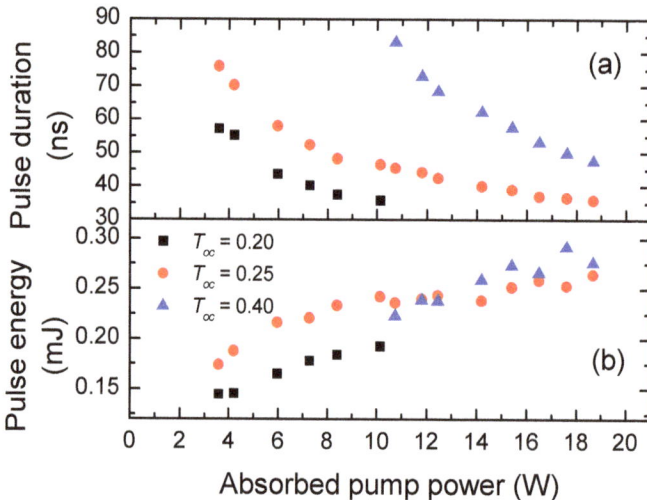

Figure 3. (a) Pulse duration and (b) energy vs. absorbed pump power.

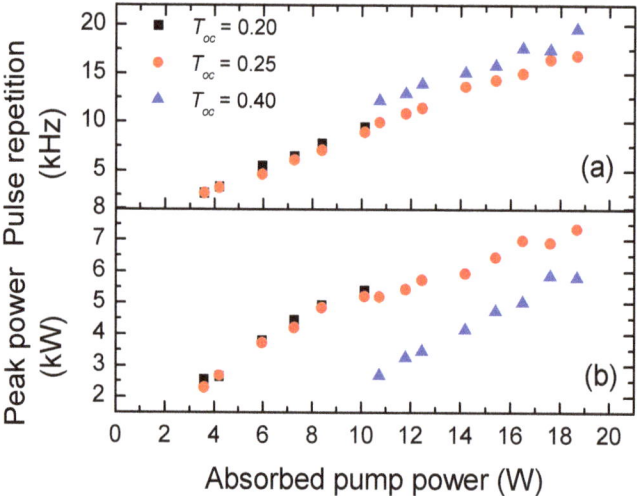

Figure 4. (a) Pulse repetition rate and (b) peak power vs. absorbed pump power.

The energy and duration of PQS lasers, and then peak power, are relative to some parameters determined by the geometry of resonator and the pump laser intensity. To improve the pulse peak power, we increase the pump laser-mode radius focused into the gain crystal to ~600 μm by replacing the focus lens F1 with the other lens with a focus length of 75 mm for the laser with the 0.25 transmittance of OC. These experimental results were shown in Figure 5, Figure 6. For comparison, the results from the laser with ~400 μm pump mode radius were simultaneously plotted in the figures. The output power and slope efficiency declined slightly due to the slight mode-mismatch between laser-mode and pump-mode (shown in Figure 5).

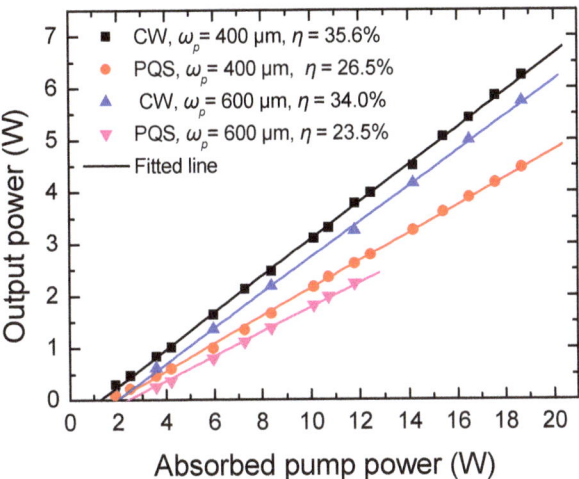

Figure 5. Output power vs. absorbed pump power (T_{oc} = 0.25).

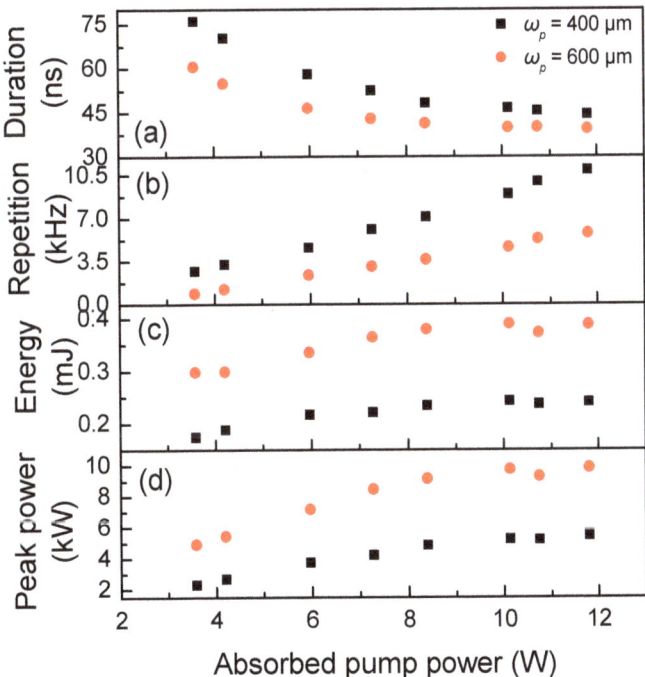

Figure 6. Pulse characteristics of the PQS laser (T_{oc} = 0.25), (**a**) duration, (**b**) repetition, (**c**) energy, and (**d**) peak power.

In addition, the maximum absorbed pump power that can exhibit stable Q-switching phenomenon did not exceed 11.8 W which may attribute to the changing of the oscillating laser-mode in the cavity due to thermal effect of laser crystal. The pulse properties of the PQS laser were shown in Figure 6. It is obvious from Figure 6a that the pulse duration is narrower than that of the laser with 400 μm pump mode radium within the total range of absorbed pump power (3.6–11.8 W). The pulse repetition was significantly low varying from 0.8 kHz to 5.7 kHz with the increase of pump power (shown in Figure 6b). The narrowest pulse duration of 39 ns was obtained at the absorbed pump power of 11.8 W with the repetition rate of 5.7 kHz. The maximum energy of 0.39 mJ was achieved at absorbed pump power of 11.8 W (shown in Figure 6c). The maximum peak power is up to 9.9 kW (shown in Figure 6d), evidently greater than that of the laser with 400 µm pump mode radium. At the maximum absorbed pump power of 11.8 W, the single pulse trace were recorded through scattering the laser into a fast PIN photodiode connected to a Lecroy digital oscilloscope (600 MHz bandwidth), as shown in Figure 7. We also measured the emission wavelength of the CW and Q-switched lasers with various output couplers. The output laser wavelength kept in about 2.05 µm.

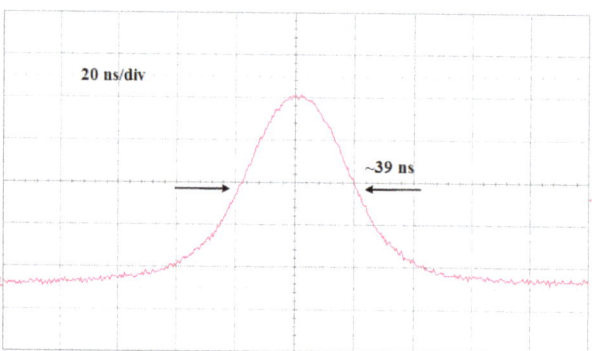

Figure 7. Single pulse trace at the maximum absorbed pump power of 11.8 W.

4. Conclusions

In brief, we have researched the properties of the diode-pumped PQS Tm, Ho: GdVO$_4$ lasers with U-shaped resonator geometry at cryogenic temperature. With the pump mode-radius of near 400 µm and OC transmittance of 0.25, the PQS laser exhibited stable Q-switching performance within the greatest range of absorbed pump power (3.5–18.7 W), though output power was less than that of the laser with OC transmittance of 0.40. At the 18.7 W absorbed pump power, the PQS laser generated the maximum peak power of ~7.3 kW with the pulse duration of ~36 ns. Increasing the pump mode-radius to ~600 µm, the repetition rate was obviously low, changing from 0.8 kHz to 5.7 kHz, in comparison with the laser with transmittance of 0.25 and 400 µm pump mode-radium, though the stable Q-switching can only be realized within the range of 3.6 W–11.8 W. The largest pulse peak power of 9.9 kW was achieved with the maximum energy of 0.39 mJ at absorbed pump power of 11.8 W. The future work will focus on narrowing the pulse duration and enhancing the pulse peak power through optimizing the resonator geometry.

Author Contributions: Conceptualization, Y.D. and B.Y., methodology, B.Y., Y.D. and T.D.; software, Y.D., J.T., X.C. and T.D.; validation, Y.D., H.S. and H.K.; formal analysis, T.D., Y.D., H.S., H.K. and H.X.; investigation, Y.D. and T.D.; resources, B.Y.; data curation, J.T. and X.C.; writing—original draft preparation, Y.D.; writing—review and editing, B.Y. and T.D.; visualization, Y.D., H.S., H.K. and H.X.; supervision, B.Y.; project administration, Y.D. and B.Y.; funding acquisition, Y.D. and B.Y. All authors have read and agreed to the published version of the manuscript.

Funding: This research was funded by National Natural Science Foundation of China, grant number 61805074, and Natural Science Foundation of Heilongjiang Province of China, grant number LH2019F034.

Data Availability Statement: The data presented in this study are available on request from the corresponding author.

Acknowledgments: This work was supported by China Postdoctoral Science Foundation (2016M601415).

Conflicts of Interest: The authors declare no conflict of interest.

References

1. Wang, Y.; Ju, Y.; Dai, T.; Yan, D.; Chen, Y.; Fang, X.; Duan, X.; Yao, B. Continuously tunable high-power single-longitudinal-mode Ho:YLF laser around the P12 CO_2 absorption line. *Opt. Lett.* **2020**, *45*, 6691–6694. [CrossRef]
2. Pyo, H.; Kim, H.; Kang, H.W. Evaluations on laser ablation of exvivo porcine stomach tissue for development of Ho:YAG-assisted endoscopic submucosal dissection (ESD). *Laser Med. Sci.* **2020**. [CrossRef]
3. Wang, R.X.; Yao, B.Q.; Zhao, B.R.; Chen, Y.; Liu, G.Y.; Dai, T.Y.; Duan, X.M. Single-longitudinal-mode Ho:YVO$_4$ MOPA system with a passively Q-switched unidirectional ring oscillator. *Opt. Express* **2019**, *27*, 34618–34625. [CrossRef] [PubMed]
4. Li, L.; Yang, X.; Yang, Y.; Zhou, L.; Wang, W.; Yu, X.; Wang, Y. A high-power, long-wavelength infrared ZnGeP$_2$ OPO pumped by a Q-switched Tm, Ho: GdVO$_4$ laser. *J. Russ. Laser Res.* **2017**, *38*, 305–310. [CrossRef]
5. Goldberg, L.; King, V.; Cole, B.; Hays, A. Passively Q-switched 10 mJ Tm:YLF laser with efficient OPO conversion to mid-IR. In Proceedings of the SPIE, San Francisco, CA, USA, 21 February 2020; p. 11259.
6. Loiko, P.A.; Yumashev, K.V.; Matrosov, V.N. and Kuleshov, N.V. Dispersion and anisotropy of thermo-optic coefficients in tetragonal GdVO$_4$ and YVO$_4$ laser host crystals. *Appl. Opt.* **2013**, *52*, 698–705. [CrossRef]
7. Nadimi, M.; Waritanant, T.; Major, A. High power and beam quality continuous-wave Nd:GdVO$_4$ laser in-band diode-pumped at 912 nm. *Photon. Res.* **2017**, *5*, 346–349. [CrossRef]
8. Ryba-Romanowski, W.; Lisiecki, R.; Jelínková, H.; Šulc, J. Thulium-doped vanadate crystals: Growth, spectroscopy and laser performance. *Prog. Quant. Electron.* **2011**, *35*, 109–157. [CrossRef]
9. Loiko, P.; Boguslawski, J.; Maria Serres, J.; Kifle, E.; Kowalczyk, M.; Mateos, X.; Sotor, J.; Zybala, R.; Mars, K.; Mikula, A.; et al. Sb$_2$Te$_3$ thin film for the passive Q-switching of a Tm:GdVO$_4$ laser. *Opt. Mate. Express* **2018**, *8*, 1723–1732. [CrossRef]
10. Duan, X.M.; Ding, Y.; Yao, B.Q.; Wang, Y.Z. High power acousto-optical Q-switched Tm:YLF-pumped Ho:GdVO$_4$ laser. *Optik* **2018**, *163*, 39–42. [CrossRef]
11. Yu, X.; Kang, J.; Zhou, L.; Xu, C.; Li, L.; Li, S.; Yang, Y. Passive Q-switched operation of a c-cut Tm, Ho:LuVO$_4$ laser with a few-layer WSe$_2$ saturable absorber. *J. Russ. Laser Res.* **2019**, *40*, 288–292. [CrossRef]
12. Li, L.; Li, T.; Zhou, L.; Fan, J.; Yang, Y.; Xie, W.; Li, S. Passively Q-switched diode-pumped Tm, Ho:LuVO$_4$ laser with a black phosphorus saturable absorber. *Chin. Phys. B* **2019**, *28*, 094205. [CrossRef]
13. Simanovskii, D.M.; Schwettman, H.A.; Lee, H.; Welch, A.J. Midinfrared optical breakdown in transparent dielectrics. *Phys. Rev. Lett.* **2003**, *91*, 107601. [CrossRef]
14. Mirov, S.; Fedorov, V.; Moskalev, I.; Martyshkin, D.; Kim, C. Progress in Cr^{2+} and Fe^{2+} doped mid-IR laser materials. *Laser Photonics Rev.* **2010**, *4*, 21–41. [CrossRef]
15. Cole, B.; Goldberg, L. Highly efficient passively Q-switched Tm:YAP laser using a Cr:ZnS saturable absorber. *Opt. Lett.* **2017**, *42*, 2259–2262. [CrossRef] [PubMed]
16. Yuan, J.H.; Yao, B.Q.; Dai, T.Y.; Gao, Y.C.; Yu, J.; Sun, J.H. High peak power, high-repetition rate passively Q-switching of a holmium ceramic laser. *Laser Phys.* **2020**, *30*, 035004. [CrossRef]
17. Zhao, B.R.; Chen, Y.; Yao, B.Q.; Cui, Z.; Bai, S.; Yang, H.Y.; Duan, X.M.; Li, J.; Shen, Y.J.; Qian, C.P.; et al. Repetition-frequency-controllable double Q-switched Ho:LuAG laser with acousto-optic modulator and Cr^{2+}:ZnS saturable absorber. *Opt. Eng.* **2016**, *55*, 4. [CrossRef]
18. Zhang, X.L.; Zhang, S.; Xiao, N.N.; Zhao, J.Q.; Li, L.; Cui, J.H. Diode-pumped passivelyQ-switched dual-wavelengthc-cut Tm, Ho:LLF laser at 2 µm. *Laser Phys. Lett.* **2014**, *11*, 035801. [CrossRef]
19. Zhang, X.L.; Bao, X.J.; Li, L.; Li, H.; Cui, J.H. Laser diode end-pumped passively Q-switched Tm, Ho:YLF laser with Cr:ZnS as a saturable absorber. *Opt. Commun.* **2012**, *285*, 2122–2127. [CrossRef]
20. Serres, J.M.; Loiko, P.; Mateos, X.; Jambunathan, V.; Yasukevich, A.S.; Yumashev, K.V.; Petrov, V.; Griebner, U.; Aguilo, M.; Diaz, F. Passive Q-switching of a Tm, Ho:KLu(WO$_4$)$_2$ microchip laser by a Cr:ZnS saturable absorber. *Appl. Opt.* **2016**, *55*, 3757–3763. [CrossRef]
21. Du, Y.Q.; Yao, B.Q.; Duan, X.M.; Cui, Z.; Ding, Y.; Ju, Y.L.; Shen, Z.C. Cr:ZnS saturable absorber passively Q-switched Tm,Ho:GdVO$_4$ laser. *Opt. Express* **2013**, *21*, 26506–26512. [CrossRef]
22. Du, Y.Q.; Yao, B.Q.; Cui, Z.; Duan, X.M.; Dai, T.Y.; Ju, Y.L.; Pan, Y.B.; Chen, M.; Shen, Z.C. Passively Q-switched Tm, Ho:YVO$_4$ Laser with Cr:ZnS Saturable Absorber at 2 µm. *Chin. Phys. Lett.* **2014**, *31*, 064209. [CrossRef]
23. Du, Y.Q.; Yao, B.Q.; Liu, W.; Cui, Z.; Duan, X.M.; Ju, Y.L.; Yu, H. Highly efficient passively Q-switched Tm, Ho:GdVO$_4$ laser with kilowatt peak power. *Opt. Eng.* **2016**, *55*, 046112. [CrossRef]

Article

The Investigation on Mid-Far Infrared Nonlinear Crystal AgGaGe₅Se₁₂ (AGGSe)

Youbao Ni [1,*], Qianqian Hu [1,2], Haixin Wu [1], Weimin Han [1,2], Xuezhou Yu [1,2] and Mingsheng Mao [1]

[1] Anhui Provincial Key Laboratory of Photonic Devices and Material, Anhui Institute of Optics and Fine Mechanics and Hefei Institute of Physical Science, Chinese Academy of Sciences, Hefei 230031, China; hqq1001@mail.ustc.edu.cn (Q.H.); hxwu@aiofm.ac.cn (H.W.); han19941023@mail.ustc.edu.cn (W.H.); u123xue@mail.ustc.edu.cn (X.Y.); CH123452021@126.com (M.M.)
[2] Graduate School, University of Science and Technology of China, Hefei 230026, China
* Correspondence: ybni@aiofm.ac.cn

Abstract: 3–5, 8–14 μm mid-far infrared (MF-IR) coherent lights generated by nonlinear optical (NLO) crystals are crucial for many industrial and military applications. AgGaGe₅Se₁₂ (AGGSe) is a promising NLO candidate because of its good optical performance. In this paper, the large AGGSe single crystal of 35 mm diameter and 80 mm length was obtained by the seed-aided Bridgman method. The crystalline quality was characterized with X-ray diffraction, rocking curve, transmission spectrum. The FWHM of the (210) peak was about 0.05° and the IR transmission was about 60% (1–10 μm, 6 mm thick). Additionally, it performed well in 8 μm frequency doubling, with a maximum output power of about 41 mW, corresponding to an optical-to-optical conversion efficiency of 3.2%. The laser induced damage threshold (LIDT) value was about 200 MW/cm² (1.06 μm, 20 ns, 1 Hz).

Keywords: nonlinear infrared optical crystal; AgGaGe₅Se₁₂ crystal; Bridgman growth method

1. Introduction

In recent years, 3–5, 8–14 μm mid-far infrared (MF-IR) laser sources have drawn increasing attention due to their potential applications in the fields of industry and the military, such as atmospheric monitoring, remote sensing, and IR countermeasures [1,2]. The nonlinear optical (NLO) frequency conversion is an effective and promising approach to obtain tunable MF-IR coherent light. Here, the NLO crystal is regarded as one of the core components in this system. At present, many MF-IR NLO crystals such as AgGaS₂, AgGaSe₂, ZnGeP₂, CdSiP₂, OP-GaAs, OP-GaP, BaGa₄S₇, and BaGa₄Se₇ have been developed and widely used [3–7].

AgGaGe₅Se₁₂ (AGGSe), as a new type of NLO crystal, was first discovered by VV Badikov in the AgGaSe₂-nGeSe₂ solid solutions systems with n = 5 [8]. AGGSe has wide spectral region (0.6–16.5 μm) and band gap (2.2 ev), large LIDT (220 MW/cm², 1.06 μm, 15 ns, 1 Hz) [9–11]. Additionally, it can achieve 3–5, 8–14 μm mid-far infrared wavelength, be pumped by commercially available 1.06 μm Nd:YAG or 0.8–1.0 μm Ti:Sapphire lasers, with the type of second harmonic generation (SHG) [11,12], optical parametric amplification (OPA) [13], and also for difference frequency generation (DFG) [14]. It may offer some advantages over commercial crystals AgGaS₂, AgGaSe₂ if high crystal quality could be achieved.

However, up to now, only a few attempts have been reported regarding the growth, fabrication and application of AGGSe crystals; probably due to the high equilibrium partial pressure and high volatility of Se, micro-cracks induced by large anisotropy thermal expansion, stoichiometric twins and variation, obtaining large-sized and high-quality AGGSe single crystals is still a considerable challenge [14–17].

In this paper, we report the recent progress in the growth of single AGGSe crystals in our laboratory. Finally, the high-quality AGGSe crystals were grown routinely with the dimension size 35 mm in diameter and 80 mm in length. The related optimized growth parameters were chosen and discussed, and the certain properties, including the X-ray diffraction, rocking curve, SHG laser experiment and LIDT were also tested.

2. Materials and Methods

2.1. Poly-Crystalline Prepared

The AGGSe poly-crystalline was prepared via a directed high temperature solid-state reaction technique. High purity elements Ag, Ga Ge, Se (EMei Semiconductor Co. Ltd., Leshan, China) with 5–9's and 6–9's grade were used as starting materials. To obtain uniform and single-phase materials, a suitable heating program was designed in our previous work [9], and 200 g AGGSe poly-crystalline could be obtained in one run. The identity of the obtained sample was tested by powder XRD analysis, and it was in excellent agreement with the calculated pattern on the basis of the single crystal crystallographic data of AGGSe, without any detectable impurity.

2.2. Crystal Growth of AGGSe

2.2.1. Seed Orientation Selected

AGGSe single crystal was grown by seed aided vertical Bridgman method in our laboratory. In [100], [110], [010] and so on, several seed orientations were applied for the crystal growth. For the AGGSe crystal, the largest nonlinear coefficient value was d_{31} in XZ plane (we used the correspondence XYZ = cab between the crystallographic and the principal optic axes [14]). It would be beneficial to grow the crystal along the maximum nonlinear coefficient plane or phase matching (PM) orientation, which is an effective way to fabricate long laser elements by increasing the conversion efficiency. For this reason, the seed orientation is focused on [100] in the process of subsequent crystal growth, and high-quality seeds are selected and are normally used repeatedly for a large number of experiments. The seeds are smaller in diameter (about 5 mm with 50 mm in length) with respect to the final diameter of the grown AGGSe crystal (about 35 mm).

2.2.2. Melting Point (or Seed Melting Point) Ascertained

Due to the opacity of the growth chamber, the AGGSe seed-melt interface during growth procedure cannot be observed. In fact, the thermal environment of the growth chamber and the melting point of crystal are vitally important, and tiny temperature fluctuations could sometimes remelt the entire length of the seed. Normally, the heat transfer between the furnace, crucible, melt and growth chamber should be optimized and fixed, and here the exact AGGSe (seed) melting point is important for the crystal growth. As reported in the reference, the melting point of AGGSe is 713 °C [8], 711 °C [14].

For a guide to crystal growth under our conditions, the melting point of AGGSe had to be further investigated and the thermogravimetric and differential thermal analysis instrument (TG-DTA, Diamond TG/DTA, Perkin-Elmer) was used. Here, about 5 mg of AGGSe crystal sample was measured by the heating rate of 5 °C/min in the temperature range of 30–800 °C. Only one absorption peak occurred at 715.2 °C in the TG-DSC curves (Figure 1).

Then, about 20 g poly-crystalline with a random seed orientation (about 30 mm long) was compacted, evacuated and sealed in a cone-shaped growth quartz. Several S-type thermocouples were fixed at the outside of the ampoule near the seed to measure the temperature accurately. Subsequently, the quartz was placed in a special modified resistance furnace with visual observation hole, which was heated to temperature near 710 °C gradually. The procedure of seed melting could been seen clearly as the temperature arise through the observation hole. Normally, the seed needs to be controlled and partially remelted ($\approx$1 cm) to expose a fresh growth interface. This process is crucial and requires adequate care and rich experience, and the seed melting point of AGGSe was assured as 714.3 °C with a potentiometer under above test condition.

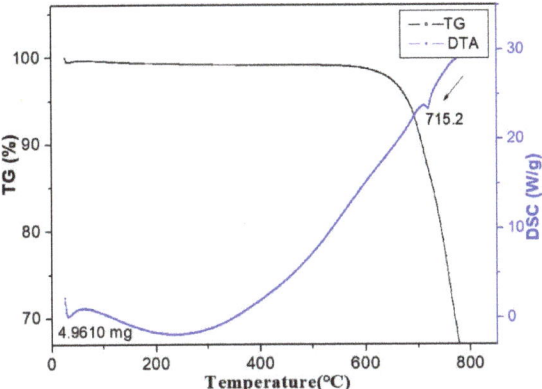

Figure 1. The TG/DSC curves of AGGSe crystal.

2.2.3. Other Growth Parameters Chosen

Different growth parameters, including the crucible pulling down speed, the temperature gradient, the rotation rate and the cooling rate, should be optimized and well controlled over the long AGGSe crystal growth period.

In fact, the growth rates greater than 6 mm per day often produced inclusion globule defects and severe cracking, twins and serious cracks were found in the AGGSe crystal when we attempted to accelerate the growth rate. To keep other experimental parameters invariable, different temperature gradients were also tested. Finally, the low growth rates ($\leq$5 mm/d) and low temperature gradients ($\leq$10 K/cm) were selected, especially in the large size crystal growth procedure, and these probably helped to avoid the thermal stresses due to large changes in diameter from the capillary to the bulk portion.

The effect of a low rotation rate had been evaluated. In the range 0–10 rpm, the rotation of quartz had no effect under the abovementioned growth conditions and the melt and quartz probably move as one body, and the interface shape remained slightly convex to the liquid regardless of the rotational conditions. The higher rotation rate was not studied because of the restrictions of the equipment.

Finally, the above factors had been considered and AGGSe crystal was grown by a seed-aided vertical Bridgman method. The crystal, 20 mm in diameter and 50 mm in length, is shown in Figure 2a. Moreover, by further growth parameter optimization, larger-sized AGGSe crystals with diameters of 30 mm and 35 mm were successfully obtained, as shown in Figure 2b,c. Some typical AGGSe crystal slices and samples 6 mm thick were prepared and are shown in Figure 3a,b. Under the table lamp light, hardly any cracks, precipitates, voids, twins, or micro bubbles could been seen in the crystal (Figures 2d and 3c).

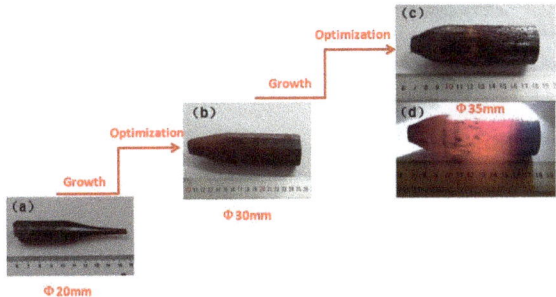

Figure 2. The AGGSe crystals with dimensions of (**a**) Φ20 mm, (**b**) Φ30 mm, (**c**) Φ35 mm and (**d**) Φ35 mm under table lamp light.

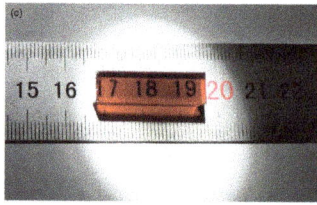

Figure 3. The prepared AGGSe crystal (**a**) slices, (**b**) sample and (**c**) sample under table lamp light with 6 mm thick.

3. Results and Discussion

3.1. Rocking Curve Measurement

The crystalline quality assessment of AGGSe was performed using a Bruker D8 ADVANCE X-ray diffractometer. A sample wafer was cut and mechanically polished on both sides for the rocking curve measurements. The shape of the peak has good symmetry without any sign of peak splitting. The intensity of the diffraction peak was high and the full widths at half maximum (FWHM) of the (210) diffraction peak was about 0.05°.

3.2. Transmission Spectrum Test

The optical transmission properties were recorded by using a UV-vis-NIR spectrophotometer (PerkinElmer Lambda 950, Waltham, MA USA) and a Fourier transform infrared (FTIR) spectrophotometer (Bruker Vertex 70, Berlin, Germany) in the range of 500–2500 nm and 4000–400 cm^{-1} respectively. The sample (Figure 3b) 6 mm thick was polished well on both sides before testing.

Figure 4a,b is the transparency spectrum of the crystal in the whole transparency range—0.63 to 2.5, 2.5 to 16.5 µm—respectively, and the crystal exhibited about 60% in 1–10 µm wavelength without obvious absorption peaks. Figure 4c exhibits the absorption coefficients (calculated according to the Beer–Lambert law with multiple reflections [18]) for the crystal and the value is about 0.06–0.1 cm^{-1} in the 1–10 µm wavelength, indicating the crystal has a fine optical quality.

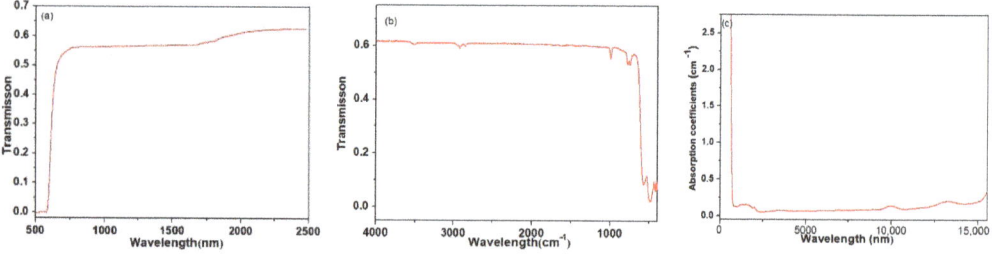

Figure 4. Transparency spectrum of AGGSe crystal (6 mm thick) (**a**) in the range of 500–2500 nm, (**b**) 4000–450 cm^{-1}, and (**c**) the optical absorption spectra in the 950–15,500 nm by calculation.

3.3. SHG Experiment

The experimental setup of the SHG of AGGSe is shown in Figure 5 [12]. Here, the 8.0 µm fundamental wave, with a pulse repetition frequency of 1 kHz, a pulse width of 27 ns, a beam quality $M_x^2 = 1.54$, $M_y^2 = 1.95$, and energy 1.3 mJ per single-pulse, was applied for the SHG experiment, which was obtained by a homemade 2.1 µm Ho, Tm:YLF laser-pumped ZnGeP$_2$ OPO procedure. The pump beam was vertically polarized to meet oo–e interaction. The uncoated AGGSe crystal with a cross face 5.5 × 6.5 mm^2 and length 10 mm, cut angle $\theta = 67.5°$, $\varphi = 0°$, xz plane, was wrapped in indium foil and fixed in a copper heat sink.

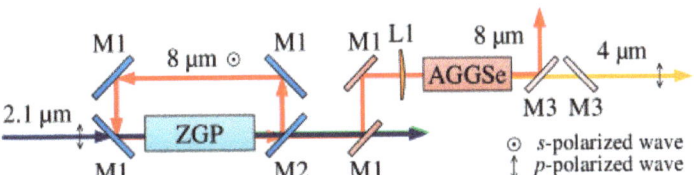

Figure 5. The Experimental setup of AGGSe SHG [12].

When the 8 µm polarized fundamental wave was injected into the AGGSe crystal, by rotating along the y axis of the crystal, the phase-matching angle θ could be changed, and a 4.0 µm laser could be obtained with a maximum output power of about 41 mW, corresponding to an optical-to-optical conversion efficiency of 3.2%. There is a deviation in the phase-matching angle θ of about 0.38° between the maximum output power (67.13°) angle and the calculated angle ($\theta = 67.5°$), which may be caused by the deviation in the crystal cutting, the errors in the Sellmeier Equation and the experimental errors.

3.4. LIDT Measurement

LIDT is one of the most important factors in the evaluation of crystal quality, especially in the use of high energy applications. Here, a standard one-on-one test procedure (ISO11254-1, 2011) was applied to test the crystal AGGSe. The sample shown in Figure 3b was used in this measurement. A commercial 1.06 µm Nd:YAG laser with pulse duration 20 ns, repetition 1 Hz and beam diameter 1.0 mm, was used as a laser source. The power of the laser increased (5 to 20 mJ) and was marked as one group in the crystal. Simultaneously the spots were observed under an optical microscope. Under four groups of experiments, the crystal surface usually damaged was under about 16 mJ per pulse power input. The surface LIDT value of crystal was calculated 200 MW/cm^2. All the above characterizations demonstrate that high-quality crystals were obtained by an optimized seed-aided Bridgman growth method.

4. Conclusions

In summary, using a seed-aided Bridgman growth method, 35 mm in diameter and 80 mm in length AGGSe crystal boule could grow integrated. The test results showed that the crystal has a high crystallinity (FWTH ≈ 0.05°), a low absorption (0.06–0.1 cm^{-1}, 1–10 µm), a fine SHG frequency conversion ability (41 mW, conversion efficiency 3.2%) and a high LIDT value (200 MW/cm^2, 20 ns, 1 Hz).

Author Contributions: Experiment and writing original draft preparation: Y.N., formal analysis: Y.N., Q.H., review and editing: H.W., W.H., X.Y., M.M. All authors have read and agreed to the published version of the manuscript.

Funding: Supported by the Special Zone of National Defense Science and Technology Innovation project.

Institutional Review Board Statement: Not applicable.

Informed Consent Statement: Not applicable.

Data Availability Statement: Not applicable.

Conflicts of Interest: The authors declare no conflict of interest.

References

1. Das, S. Broadly tunable multi-output coherent source based on optical parametric oscillator. *Opt. Laser Technol.* **2015**, *71*, 63–67. [CrossRef]
2. Parasyuk, O.V.; Babizhetskyy, V.S.; Khyzhun, O.Y.; Levytskyy, V.O.; Kityk, I.V.; Myronchuk, G.L.; Tsisar, O.V.; Piskach, L.V.; Jedryka, J.; Maciag, A.; et al. Novel quaternary TlGaSn$_2$Se$_6$ single crystal as promising material for laser operated infrared nonlinear optical modulators. *Crystals* **2017**, *7*, 341. [CrossRef]
3. Murray, R.T.; Chandran, A.M.; Battle, R.A.; Runcorn, T.H.; Schunemann, P.G.; Zawilski, K.T.; Guha, S.; Taylor, J.R. Seeded optical parametric generation in CdSiP$_2$ pumped by a Raman fiber amplifier at 1.24 µm. *Opt. Lett.* **2021**, *46*, 2039–2042. [CrossRef] [PubMed]
4. Fu, Q.; Wu, Y.; Liang, S.; Shardlow, P.C.; Shepherd, D.P.; Alam, S.U.; Xu, L.; Richardson, D.J. Controllable duration and repetition-rate picosecond pulses from a high-average-power OP-GaAs OPO. *Opt. Express* **2020**, *28*, 32540–32548. [CrossRef] [PubMed]
5. Yao, J.; Mei, D.; Bai, L.; Lin, Z.; Yin, W.; Fu, P.; Wu, Y. BaGa$_4$Se$_7$: A New Congruent-Melting IR Nonlinear Optical Material. *Inorg. Chem.* **2010**, *49*, 9212–9216. [CrossRef] [PubMed]
6. Nikogosyan, D.N. *Nonlinear Optical Crystal: A Complete Survey*; Springer: Berlin, Germany, 2005; pp. 75–108.
7. Lei, Z.; Okunev, A.; Zhu, C.; Verozubova, G.; Yang, C. Low-angle boundaries in ZnGeP$_2$ single crystals. *J. Appl. Crystallogr.* **2018**, *51*, 361–367. [CrossRef]
8. Badikov, V.V.; Tyulyupa, A.G.; Shevyrdyaeva, G.S.; Sheina, S.G. Solid solutions in the AgGaS$_2$-GeS$_2$ and AgGaSe$_2$-GeSe$_2$ systems. *Inorg. Mater.* **1991**, *27*, 177–180.
9. Ni, Y.; Wu, H.; Xiao, R.; Huang, C.; Wang, Z.; Mao, M.; Qi, M.; Cheng, G. Growth and optical properties of single AgGaGe$_5$Se$_{12}$ (AGGSe) crystal. *Opt. Mater.* **2015**, *42*, 458–461. [CrossRef]
10. Huang, W.; Wu, J.; Chen, B.; Li, J.; He, Z. Crystal growth and thermal annealing of AgGaGe$_5$Se$_{12}$ crystal. *J. Alloy. Compd.* **2021**, *862*, 158002. [CrossRef]
11. Reshak, A.H.; Parasyuk, O.V.; Fedorchuk, A.O.; Kamarudin, H.; Auluck, S.; Chyský, J. Optical Spectra and Band Structure of Ag$_x$Ga$_x$Ge$_{1-x}$Se$_2$ (x = 0.333, 0.250, 0.200, 0.167) Single Crystals: Experiment and Theory. *J. Phys. Chem. B* **2013**, *117*, 15220–15231. [CrossRef]
12. Chen, Y.; Yao, B.; Wu, H.; Ni, Y.; Liu, G.; Dai, T.; Duan, X.; Ju, Y. Broadband second-harmonic and sum-frequency generation with a long-wave infrared laser in AgGaGe$_5$Se$_{12}$. *Appl. Opt.* **2020**, *59*, 5247–5251. [CrossRef] [PubMed]
13. Knuteson, D.; Singh, N.; Kanner, G.; Berghmans, A.; Wagner, B.; Kahler, D.; McLaughlin, S.; Suhre, D.; Gottlieb, M. Quaternary AgGaGenSe$_{2(n+1)}$ crystals for NLO applications. *J. Cryst. Growth* **2010**, *312*, 1114–1117. [CrossRef]
14. Petrov, V.; Noack, F.; Badikov, V.; Shevyrdyaeva, G.; Panyutin, V.; Chizhikov, V. Phase-matching and femtosecond difference-frequency generation in the quaternary semiconductor AgGaGe$_5$Se$_{12}$. *Appl. Opt.* **2004**, *43*, 4590–4597. [CrossRef] [PubMed]
15. Schunemann, P.G.; Zawilski, K.T.; Pollak, T.M. Horizontal gradient freeze growth of AgGaGeS$_4$ and AgGaGe$_5$Se$_{12}$. *J. Cryst. Growth* **2006**, *287*, 248–251. [CrossRef]
16. Singh, N.B.; Knuteson, D.J.; Kanner, G.; Berghmans, A.; Green, K.; Wagner, B.; Kahler, D.; King, M.; McLaughlin, S. Ternary and quaternary selenide crystals for nonlinear optical applications. *Proc. SPIE* **2011**, *8120*, 812002.
17. Zhao, B.J.; Zhu, S.F.; He, Z.Y.; Chen, B.J. Research progress of middle and far infrared nonlinear optical crystals AgGaGenQ$_{2(n+1)}$(Q=S, Se). *J. Synth. Cryst.* **2020**, *49*, 1417–1426.
18. Tochitsky, S.Y.; Petukhov, V.O.; Gorobets, V.A.; Churakov, V.V.; Jakimovich, V.N. Efficient continuous-wave frequency doubling of a tunable CO$_2$ laser in AgGaSe$_2$. *Appl. Opt.* **1997**, *36*, 1882–1888. [CrossRef] [PubMed]

Article

A High-Energy, Narrow-Pulse-Width, Long-Wave Infrared Laser Based on ZGP Crystal

Chuanpeng Qian [1,2], Ting Yu [1,2,3,*], Jing Liu [1,2], Yuyao Jiang [1,2,4], Sijie Wang [1,2,3], Xiangchun Shi [1,2,3], Xisheng Ye [1,2,3,*] and Weibiao Chen [1,2,3]

[1] Laboratory of High Power Fiber Laser Technology, Shanghai Institute of Optics and Fine Mechanics, Chinese Academy of Sciences, Shanghai 201800, China; cpqian@siom.ac.cn (C.Q.); liujing@siom.ac.cn (J.L.); jiangyuyao@shu.edu.cn (Y.J.); wangsijie@siom.ac.cn (S.W.); shixc@siom.ac.cn (X.S.); wbchen@siom.ac.cn (W.C.)
[2] Shanghai Key Laboratory of All Solid-State Laser and Applied Techniques, Shanghai Institute of Optics and Fine Mechanics, Chinese Academy of Sciences, Shanghai 201800, China
[3] College of Materials Science and Opto-Electronic Technology, University of Chinese Academy of Sciences, Beijing 100049, China
[4] College of Science, Shanghai University, Shanghai 200444, China
* Correspondence: yuting@siom.ac.cn (T.Y.); xsye@siom.ac.cn (X.Y.)

Abstract: In this paper, we present a high-energy, narrow pulse-width, long-wave infrared laser based on a $ZnGeP_2$ (ZGP) optical parametric oscillator (OPO). The pump source is a 2.1 μm three-stage Ho:YAG master oscillator power-amplifier (MOPA). At a repetition frequency of 1 kHz, the Ho:YAG MOPA system outputs the maximal average power of 52.1 W, which corresponds to the shortest pulse width of 14.40 ns. By using the Ho:YAG MOPA system as the pump source, the maximal average output powers of 3.15 W at 8.2 μm and 11.4 W at 2.8 μm were achieved in a ZGP OPO. The peak wavelength and linewidth (FWHM) of the long-wave infrared laser were 8156 nm and 270 nm, respectively. At the maximal output level, the pulse width and beam quality factor M^2 were measured to be 8.10 ns and 6.2, respectively.

Keywords: long-wave infrared; $ZnGeP_2$ crystal; Ho:YAG MOPA

1. Introduction

As an important atmospheric transmission window, long-wave infrared lasers (8–12 μm) have been extensively applied in many fields, such as lidar, spectroscopy, and national defense [1,2]. Among the many ways to obtain a long-wave infrared laser, the optical parametric oscillator is an attractive approach due to its wide wavelength-tuning range and high conversion efficiency [3]. As the core component, the characteristics of nonlinear optical crystals determine the performance of nonlinear frequency conversion. At present, nonlinear crystals suitable for generating a long-wave infrared laser mainly include OP-GaAs, $AgGaSe_2$, CdSe, $BaGa_4Se_7$, and $ZnGeP_2$ (ZGP).

The nonlinear coefficient of OP-GaAs is very large (d_{14} = 94 pm/V), and it was used to achieve an average pulse energy of 0.18 mJ at 8.5 μm [4] and 16.2 μJ at 10.6 μm [5], corresponding to the repetition frequencies of 2 and 50 kHz, respectively. $AgGaSe_2$ has a low damage threshold (18 MW/cm^2), which limits its ability to obtain a large-energy long-wave infrared laser. The highest energy of a long-wave infrared laser by $AgGaSe_2$ was about hundreds of microjoules [6–8]. CdSe has a small nonlinear coefficient (d_{31} = 18 pm/V) and a moderate damage threshold (56 MW/cm^2). Due to its weak walk-off effect, the disadvantage of a small nonlinear coefficient can be compensated by increasing the crystal length. The pulse energy of a long-wave infrared laser achieved 1.05 mJ at 10.1 μm [9] and 0.8 mJ at 11 μm [10]. $BaGa_4Se_7$ has a very large damage threshold (557 MW/cm^2) and an acceptable nonlinear coefficient (d_{11} = 24.3 pm/V), but its low thermal conductivity makes it unsuitable for obtaining a high-power long-wave infrared laser. In 2018, Zhao et al.

obtained an average pulse energy of 0.31 mJ at 8.92 μm [11] with a repetition frequency of 1 kHz. The nonlinear coefficient of a ZGP crystal is high (d_{14} = 75 pm/V), and its thermal conductivity and damage threshold perform well among these nonlinear crystals. Using a ZGP crystal, the largest pulse energy of 45 mJ at 8.0 μm was achieved [12]. This result was achieved under a low repetition frequency (1 Hz). At a kilohertz frequency, the ZGP crystal obtained pulse energy of 1.26 mJ at 8.2 μm [13] and 0.35 mJ at 9.8 μm [14], with a repetition frequency of 10 kHz.

We demonstrate a high-energy, narrow-pulse-width, long-wave infrared laser with repetition frequency of 1 kHz based on a ZGP crystal. The pump source is a 2.1 μm Ho:YAG MOPA laser that can output a highest average energy of 52.1 mJ at 1 kHz. The pulse width and beam factor M^2 were measured to be 14.40 ns and 1.3, respectively. Then, a ZGP OPO with a four-mirror ring-cavity structure was used. The output energy of the long-wave infrared ZGP OPO was 3.15 mJ for the idler and 11.4 mJ for the signal. The pulse width and beam factor M^2 were measured to be 8.10 ns and 6.2 at the maximal output level.

2. Experimental Setup

The experimental setup of the Ho:YAG MOPA system is shown in Figure 1, which contains a Q-switched Ho:YAG oscillator and a three-stage Ho:YAG MOPA system. The acousto-optical modulator (AOM) Q-switched Ho:YAG oscillator had a compact L-shaped structure that consisted of an input mirror (M1) with antireflection (AR) for 1.9 μm and high-reflection (HR) for 2.1 μm, an output coupler (M2) with transmittance of 60% at 2.1 μm, and a thin-film polarizer (M3) with AR for s-polarized 2.1 μm and HT for p-polarized 2.1 μm. Its physical-cavity length was 150 mm. The Ho:YAG crystal in the oscillator with dopant concentration of 0.5%, diameter of 5 mm, and length of 30 mm was single-ended pumped by a Tm:YLF laser with the $1/e^2$ beam-waist diameter of 0.34 mm and maximal power of 28 W.

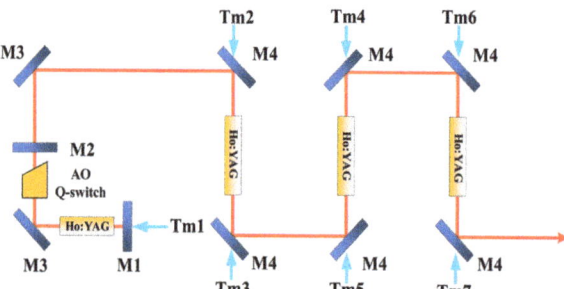

Figure 1. Experimental setup of three-stage Ho:YAG MOPA system.

The three-stage Ho:YAG MOPA system was designed and operated at a PRF of 1 kHz. The three Ho:YAG crystals with a dopant concentration of 0.3%, diameter of 5 mm, and length of 70 mm were dual-ended pumped by six Tm:YLF lasers. The output powers of the six Tm:YLF lasers were about 40 W. The beam diameters of the pump in the three Ho:YAG crystals were 0.83, 1.4, and 1.9 mm, respectively.

The experimental setup of ZGP OPO is shown in Figure 2. The ZGP OPO had a four-mirror ring cavity structure, including two mirrors (M5) with an AR p-polarized pump laser and HR idler laser, a mirror (M6) with an AR pump and signal laser and an HR idler laser, and an output coupler (M7) with transmittance of 45% for the idler laser. In order to more accurately measure the power of the idler and the signal laser two M8 mirrors and one M7 mirror were used. The oscillator resonated with the single idler laser, and the ambient humidity of the experiment facility was kept at about 10%. Both measures were designed to avoid damaging the coating of the ZGP crystal caused by water absorption around the wavelength of the signal laser. The physical length of the ring cavity was about

118 mm. With the ZGP crystal, the wavelength of the OPO was continuously tunable from 3.8 to 12.4 µm [15]. However, the optical-to-optical efficiency of the OPO decreased with the increase in wavelength. In this experiment, we adjusted the wavelength of the idler laser to 8.2 µm. The ZGP crystal had an aperture of 6 mm × 6 mm and a length of 30 mm, cut at an angle of θ = 50.8° with respect to Type I phase matching. Both ends of the ZGP crystal (School of Chemical Engineering and Technology, HIT, Ha'erbin, China) were coated with HT for the pump, signal, and idler laser. The absorption coefficients of the ZGP crystal at the 2.1 µm pump laser and 8.2 µm idler laser were measured to be 0.03 and 0.01 cm^{-1}, respectively. The pump laser from the Ho:YAG MOPA system was focused onto the ZGP crystal with a $1/e^2$ beam diameter of 3.6 mm. The crystals of the entire experimental apparatus, including Tm:YLF, Ho:YAG, and ZGP, were all wrapped in indium foil and installed into copper blocks that were cooled by the chiller. The temperatures of the Tm:YLF, Ho:YAG, and ZGP crystals were controlled to be 16, 18, and 20 °C, respectively.

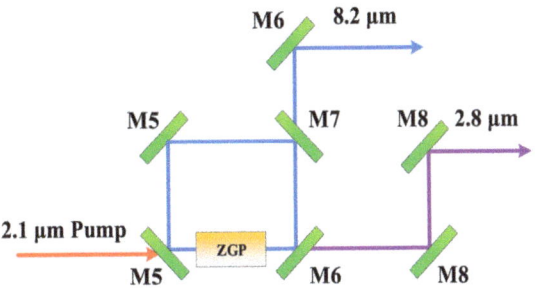

Figure 2. Experimental setup of ZGP OPO.

3. Results and Discussion

In this experiment, the output powers were measured by the same power meter (Ophir PM 150). The output performances of the Ho:YAG oscillator are shown in Figure 3. The threshold pump power of the Ho:YAG oscillator was about 8 W. Pump power of 28.0 W and a maximal average output power of 5.42 W were achieved, corresponding to the optical-to-optical conversion efficiency of 19.4%. The optical-to-optical conversion efficiency was low because we used a small mold volume to achieve the 2.1 µm narrow-pulse-width laser output. Using an InGaAs detector and a 1 GHz digital oscilloscope (Tektronix DPO4102B), we measured the minimal full-width half maximum (FWHM) of the pulse profile to be 11.56 ns, which is shown in Figure 3a. Figure 3b shows the beam-quality factor M^2 of the oscillator that was measured by the 90/10 knife-edge method. Under the maximal output condition of the Ho:YAG oscillator, the beam quality factor M^2 in the x and y directions was 1.19 and 1.26, respectively.

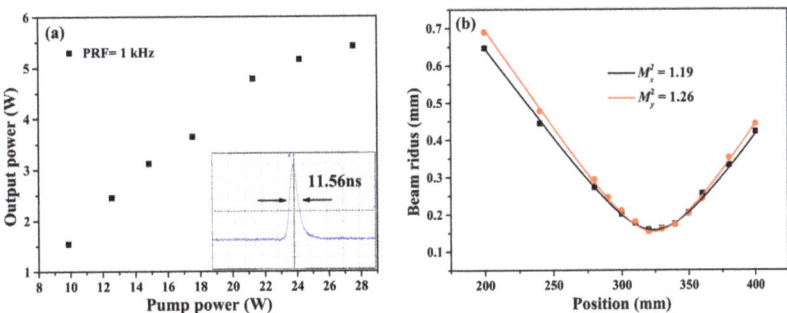

Figure 3. (a) Output power and pulse width, and (b) beam quality factor of Ho:YAG oscillator.

The 2.1 µm laser produced by the oscillator was injected into the Ho:YAG crystal in the primary amplifier after being transformed by a set of coupling lenses, as shown in Figure 4a. The output powers for each amplifier stage were 18.14, 37.5, and 52.4 W, corresponding to slope efficiencies of 24.7%, 36.0%, and 33.4%, respectively. When the amplifier moved from the first stage to the third, the pulse width of the 2.1 µm laser slightly increased, which was measured to be 12.72, 13.38, and 14.40 ns. Compared to the Ho:YAG oscillator, the beam quality factor M^2 of the Ho:YAG amplifier very slightly deteriorated, with 1.20 and 1.28 for the x and y directions, respectively. The final pulse width and beam quality of the Ho:YAG MOPA system are shown in Figure 4b,c respectively.

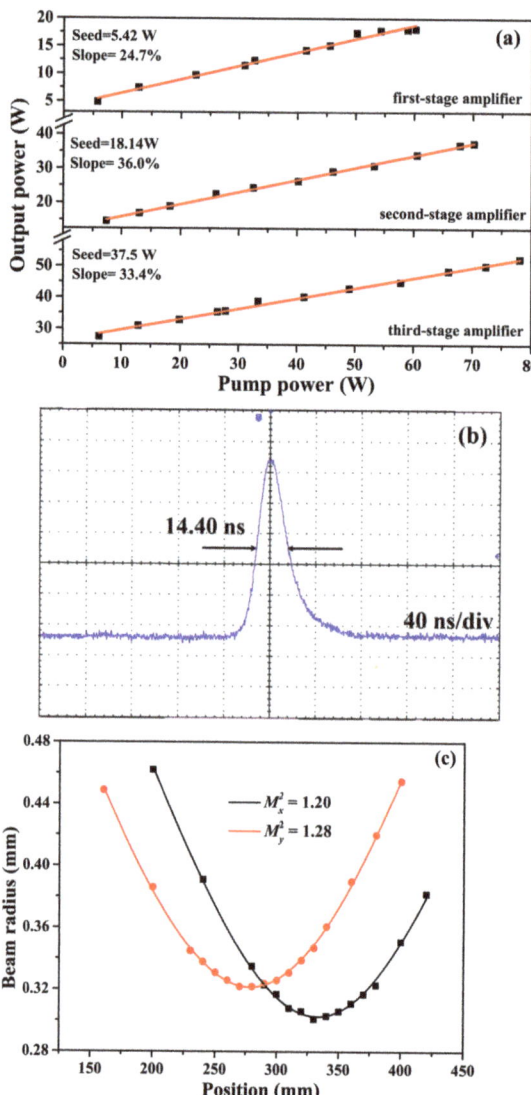

Figure 4. (**a**) Output powers, (**b**) pulse width, and (**c**) beam quality factor M^2 of three-stage Ho:YAG amplifier.

As shown in Figure 5a, the average output power of ZGP OPO was measured with an incident pump power of 52.4 W. The pump laser was injected into the crystal in a divergent way to avoid damaging the end face with the thermal-lens effect. The divergence angle was about 6 mrad. During the experiment, we gradually reduced the size of the pump spot to obtain the highest pulse energy of the long-wave infrared laser. Lastly, the beam diameter at the front-end face of the ZGP crystal was ~3.6 mm. Threshold pump power was about 21.8 W and the maximal average output power of the ZGP OPO was about 3.15 W at 8.2 μm and 11.4 W at 2.8 μm, corresponding to the slope efficiency of about 10.1% and 37.0%. The beam quality factor M^2 was measured and calculated to be 6.2 at the maximal output power, which is shown in Figure 5b.

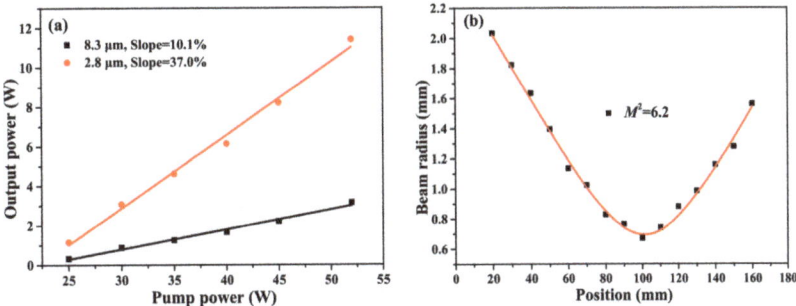

Figure 5. (a) Output power, and (b) beam quality factor M^2 of ZGP OPO.

In the experimental process, the damage threshold of the ZGP crystal had great correlation with the repetition frequency of the pump laser. In our previous work, which used a 3 kHz Ho:YAG MOPA system to pump ZGP OPO [16], the ZGP crystal was damaged when the pump power was ~73 W and the spot radius was 1.28 mm, corresponding to peak power density of 54.9 MW/cm^2. We also measured the damage threshold of the ZGP crystal at 10 kHz repetition frequency, and it was about 25.7 MW/cm^2. However, under the condition of 1 kHz repetition frequency, the ZGP crystal remained undamaged when the peak power density of the pump reached 60 MW/cm^2. For the same ZGP crystal, the damage threshold increased by more than two times under the same heat-dissipation conditions as the repetition frequency of the pump laser decreased from 10 to 1 kHz. This phenomenon could have been related to the time during which the laser was acting on the coating film. At a high repetition rate, a longer treatment time led to a higher film temperature, and this made the coating film of the ZGP crystal more vulnerable to damage.

Because the InGaAs detector could not respond to a long-wave infrared laser, we employed an HgCdTe detector combined with a signal amplifier to measure the pulse width of the 8.2 μm idler laser, which is shown in Figure 6a. The FWHM pulse width was 8.10 ns with the peak power of 0.39 MW. Using a monochromator spectrograph (Zolix, omni-λ 300i), the idler spectrum was measured and is shown in Figure 6b. Peak wavelength was 8156 nm. The corresponding linewidth (FWHM) was approximately 270 nm.

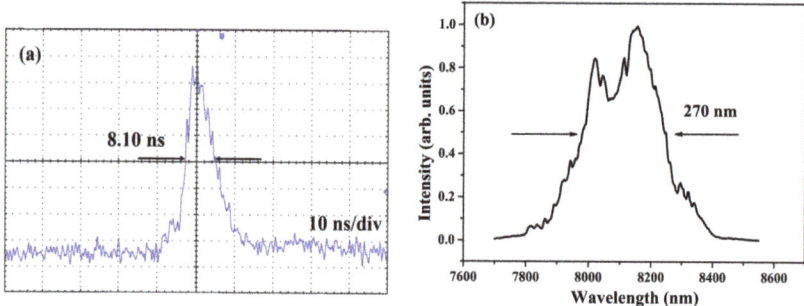

Figure 6. (a) Pulse width, (b) ZGP OPO spectrum.

4. Conclusions

The Ho:YAG oscillator was Q-switched at 1 kHz, and the pulse width was ~12 ns. The successive three-stage Ho:YAG amplifier stages increased the maximal average output power up to 52.1 W with a pulse width of ~14 ns, corresponding to the beam quality factors M^2 of 1.20 and 1.28 for the horizontal and vertical directions, respectively. With the above Ho:YAG MOPA system, maximal output powers of 3.15 W at 8.2 μm and 11.4 W at 2.8 μm were produced in ZGP OPO, with an idler laser output energy of 3.15 mJ and a pulse width of 8.10 ns. The linewidth of the long-wave infrared laser was 270 nm at a central wavelength of 8156 nm. Its beam quality factors M^2 were 6.2. As far as we know, this is the largest reported amount of pulse energy in a long-wave infrared laser at a kilohertz repetition-frequency band.

Author Contributions: Experiment and writing original draft preparation: C.Q., Y.J. and S.W.; formal analysis: T.Y., J.L. and X.S.; review and editing: X.Y. and W.C. All authors have read and agreed to the published version of the manuscript.

Funding: This work was supported by the National Natural Science Foundation of China (NSFC) (62005300).

Institutional Review Board Statement: Not applicable.

Informed Consent Statement: Not applicable.

Data Availability Statement: Not applicable.

Conflicts of Interest: The authors declare no conflict of interest.

References

1. Webber, M.E.; Pushkarsky, M.; Patel, C.K.N. Optical detection of chemical warfare agents and toxic industrial chemicals: Simulation. *J. Appl. Phys.* **2005**, *97*, 113101. [CrossRef]
2. Petrov, V. Frequency down-conversion of solid-state laser sources to the mid-infrared spectral range using non-oxide nonlinear crystals. *Prog. Quantum Electron.* **2015**, *42*, 1–106. [CrossRef]
3. Hildenbrand, A.; Kieleck, C.; Tyazhev, A.; Marchev, G.; Stöppler, G.; Eichhorn, M.; Schunemann, P.G.; Panyutin, V.L.; Petrov, V. Laser damage of the nonlinear crystals CdSiP$_2$ and ZnGeP$_2$ studied with nanosecond pulses at 1064 and 2090 nm. *Opt. Eng.* **2014**, *53*, 122511. [CrossRef]
4. Gutty, F.; Grisard, A.; Larat, C.; Papillon, D.; Schwarz, M.; Gerard, B.; Ostendorf, R.; Rattunde, M.; Wagner, J.; Lallier, E. 140 W peak power laser system tunable in the LWIR. *Opt. Express* **2017**, *25*, 18897–18906. [CrossRef] [PubMed]
5. Wueppen, J.; Nyga, S.; Jungbluth, B.; Hoffmann, D. 1.95 μm-pumped OP-GaAs optical parametric oscillator with 10.6 μm idler wavelength. *Opt. Lett.* **2016**, *41*, 4225–4228. [CrossRef] [PubMed]
6. Boyko, A.A.; Kostyukova, N.Y.; Marchev, G.M.; Pasiskevicius, V.; Kolker, D.B.; Zukauskas, A.; Petrov, V. Rb:PPKTP optical parametric oscillator with intracavity difference-frequency generation in AgGaSe$_2$. *Opt. Lett.* **2016**, *41*, 2791–2794. [CrossRef] [PubMed]
7. Boyko, A.A.; Marchev, G.M.; Petrov, V.; Pasiskevicius, V.; Kolker, D.B.; Zukauskas, A.; Kostyukova, N.Y. Intracavity-pumped, cascaded AgGaSe$_2$ optical parametric oscillator tunable from 5.8 to 18 μm. *Opt. Express* **2015**, *23*, 33460–33465. [CrossRef] [PubMed]

8. Gerhards, M. High energy and narrow bandwidth mid IR nanosecond laser system. *Opt. Commun.* **2004**, *241*, 493–497. [CrossRef]
9. Chen, Y.; Liu, G.Y.; Yang, C.; Yao, B.Q.; Wang, R.X.; Mi, S.Y.; Yang, K.; Dai, T.Y.; Duan, X.M.; Ju, Y.L. 1 W, 10.1 μm, CdSe optical parametric oscillator with continuous-wave seed injection. *Opt. Lett.* **2020**, *45*, 2119–2122. [CrossRef] [PubMed]
10. Chen, Y.; Yang, C.; Liu, G.Y.; Yao, B.Q.; Wang, R.X.; Yang, K.; Mi, S.Y.; Dai, T.Y.; Duan, X.M.; You, Y.L. 11 μm, high beam quality idler-resonant CdSe optical parametric oscillator with continuous-wave injection-seeded at 2.58 μm. *Opt. Express* **2020**, *28*, 17056–17063. [CrossRef] [PubMed]
11. Zhao, B.R.; Chen, Y.; Yao, B.Q.; Yao, J.Y.; Guo, Y.W.; Wang, R.X.; Dai, T.Y.; Duan, X.M. High-efficiency, tunable 8-9 μm $BaGa_4Se_7$ optical parametric oscillator pumped at 2.1 μm. *Opt. Mater. Express* **2018**, *8*, 3332–3337. [CrossRef]
12. Bakkland, A.; Fonnum, H.; Lippert, E.; Haakestad, M.W. Long-wave infrared source with 45 mJ pulse energy based on nonlinear conversion in $ZnGeP_2$. *Conf. Lasers Electro Opt.* **2016**. [CrossRef]
13. Liu, G.Y.; Chen, Y.; Yao, B.Q.; Yang, K.; Qian, C.P.; Dai, T.Y.; Duan, X.M. Study on long-wave infrared $ZnGeP_2$ subsequent optical parametric amplifiers with different types of phase matching of $ZnGeP_2$ crystals. *Appl. Phys. B* **2019**, *125*, 233. [CrossRef]
14. Liu, G.Y.; Chen, Y.; Yao, B.Q.; Wang, R.X.; Yang, K.; Yang, C.; Mi, S.Y.; Dai, T.Y.; Duan, X.M. 3.5 W long-wave infrared $ZnGeP_2$ optical parametric oscillator at 9.8 μm. *Opt. Lett.* **2020**, *45*, 2347–2350. [CrossRef] [PubMed]
15. Vodopyanov, K.L.; Ganikhanov, F.; Maffetone, J.P.; Zwieback, I.; Ruderman, W. $ZnGeP_2$ optical parametric oscillator with 3.8–12.4 μm tunability. *Opt. Lett.* **2000**, *25*, 841–843. [CrossRef] [PubMed]
16. Qian, C.P.; Yu, T.; Liu, J.; Jiang, Y.Y.; Wang, S.J.; Shi, X.C.; Ye, X.S.; Chen, W.B. 5.4 W, 9.4 ns pulse width, long-wave infrared ZGP OPO pumped by Ho:YAG MOPA system. *IEEE Photonics J.* **2021**, *17*. in press.

Article

High-Efficiency Ho:YAP Pulse Laser Pumped at 1989 nm

Chao Niu [1], Yan Jiang [1], Ya Wen [1], Lu Zhao [1], Xinyu Chen [1], Chunting Wu [1,*] and Tongyu Dai [2]

[1] Jilin Key Laboratory of Solid-State Laser Technology and Application, Changchun University of Science and Technology, Changchun 130022, China; leoniuc@163.com (C.N.); moyudaidai@yeah.net (Y.J.); winvene@163.com (Y.W.); sasslh@163.com (L.Z.); chenxinyucust@163.com (X.C.)

[2] National Key Laboratory of Tunable Laser Technology, Harbin Institute of Technology, Harbin 150001, China; daitongyu2006@126.com

* Correspondence: bigsnow@cust.edu.cn

Abstract: A Tm:YAP laser with an output wavelength of 1989 nm was selected for the first time as the pump source of a Q-switched Ho:YAP laser. When the absorbed power was 30 W, an average power of 18.02 W with the pulse width of 104.2 ns acousto-optic (AO) Q-switched Ho:YAP laser was obtained at a repetition frequency of 10 kHz. The slope efficiency was 70.11%, and the optical-optical conversion efficiency was 43.03%. The output center wavelength was 2129.22 nm with the line width of 0.74 nm.

Keywords: 1989 nm; Ho:YAP; AO Q-switched laser

1. Introduction

2 μm holmium (Ho^{3+}) doped solid-state lasers have important application prospects in the fields of laser ranging, laser medical treatment, environmental monitoring, and optical communication due to its near-infrared window and safety to the human eye [1–8]. In addition, 2 μm Ho^{3+} doped lasers were considered as good pump sources for mid-far infrared optical parametric oscillator (OPO) [9]. Compared with Ho:YAG, Ho:YLF and Ho:$GdVO_4$ crystals, Ho:YAP crystal has obvious advantages, such as wide absorption line, large absorption cross-section, anisotropy, short growth period, the output power was not easy to saturate, and so on. There are many absorption peaks in Ho:YAP crystal at 1.9 μm. For an a-cut Ho:YAP crystal, the absorption peaks included 1872 nm, 1907 nm, 1931 nm, 1970 nm and 2045 nm. For a b-cut Ho:YAP crystal, the absorption peaks included 1884 nm, 1923 nm, 1946 nm, 1984 nm, 2023 nm and 2059 nm [10]. For a c-cut Ho:YAP crystal, the absorption peaks included 1915, 1941, 1980, and 1996 nm [11]. A maximum absorption peak for a-, b-, and c-cut Ho:YAP crystals was about 1976 nm [12].

In recent years, there are many reports on Ho:YAP lasers. In 2009, a Tm:YLF laser with the output wavelength of 1900 nm was used to pump the continuous wave Ho:YAP laser, was reported by Duan et al. [13]. The output power was 10.2 W, with the slope efficiencies of 64.0%, the optical-optical conversion efficiencies of 52.6%, and the output wavelength of 2118 nm. In 2011, a Tm:YLF laser with output wavelength of 1910 nm was used to pump the Ho:YAP (b-cut) Q-switched laser at room temperature, was reported by Yang et al. [14]. When the Q-switched repetition frequency was 5 kHz, the output power was 18.1 W, the slope efficiencies was 45.9%, the optical-optical conversion efficiencies was 36.5%, and the output wavelength was 2118 nm. In 2012, the theoretical and experimental analysis of a Ho:YAP (a-cut) crystal of 2 μm laser was reported by Yang et al. [15]. The pump wavelength was 1900 nm. The CW output power was 15.6 W. The slope efficiencies was 63.7%, the optical-optical conversion efficiencies was 54.5%, and the output wavelength was 2118 nm. In 2012, a Tm:YLF laser with an output wavelength of 1910 nm was used to pump the Q-switched Ho:YAP (a-cut) ring laser, was reported by Dai et al. [16]. When the Q-switched repetition frequency was 1 kHz, the output power of 10.17 W was obtained, the slope efficiencies was 60%, the optical-optical conversion efficiencies was 29.5%, and the output

wavelength was 2119 nm. In 2014, a Tm fiber laser with the output wavelength of 1910 nm was used to pump the Q-switched Ho:YAP (a-cut) laser, was reported by Wang et al. [17]. When the Q-switched repetition was 10 kHz, the output power of 11.0 W was obtained with the slope efficiencies of 62.1%, the optical-optical conversion efficiencies of 26.3%, and the output wavelength of 2118.0 nm. In 2014, Tm:YLF laser with an output wavelength of 1910 nm was used to pump Ho:YAP (a-cut) Q-switched laser, was reported by Duan et al. [18]. When the Q-switched repetition frequency was 10 kHz, 17.2 W output power was obtained, the slope efficiencies was 63.2%, the optical-optical conversion efficiencies was 29%, and the output wavelength was 2118 nm. In 2016, a Tm fiber laser with the output wavelength of 1910 nm was used to pump a mode-locked Ho:YAP laser, was reported by Duan et al. [19]. The output power of 2.87 W was obtained, with a slope efficiency of 15%, an optical–optical conversion efficiency of 11.9%, and an output wavelength of 2118 nm. In 2017, a Tm fiber laser with an output wavelength of 1941 nm was used to pump a Ho:YAP (c-cut) laser, as reported by Ting et al. [11]. The output power of 29 W was obtained with the slope efficiency of 42.8%, the optical–optical conversion efficiency of 60.67%, and the output wavelength of 2118 nm. In 2018, a Tm fiber laser with an output wavelength of 1910 nm was used to pump a Ho:YAP (b-cut) laser, as reported by Duan et al. [20]. The output power was 10.5 W, the slope efficiency was 53.2%, the optical–optical conversion efficiency was 41%, and the output wavelength was 2115 nm. In 2020, a Tm:YAP laser with output wavelength of 1940 nm was used to pump the electro-optic Q-switched Ho:YAP (a-cut) laser reported by Lei et al. [21]. When the repetition frequency was 4 kHz, the output power was 6.5 W, the slope efficiency was 50.6%, the optical-to-optical conversion efficiency was 28%, and the output wavelength was 2118 nm.

As mentioned above, Tm-doped solid-state lasers or Tm fiber lasers with output wavelength of 1900 nm, 1910 nm, or 1940 nm are often used as the pumping sources of Ho:YAP lasers. Although there was no pump source whose wavelength matches the strongest absorption peak of Ho:YAP crystal, the slope efficiency of Ho:YAP lasers was quite high under all kinds of situations, such as continuous or Q-switch or mode-locked operation, which means that the Ho:YAP crystal was one of the most promising Ho^{3+} doped lasers.

Under the premise of ensuring that the Ho:YAP crystal absorbs enough pumping power, the closer the wavelength of output laser and pumping laser, the smaller the quantum loss. However, there is not report on a Ho:YAP laser pumped by a 1989 nm laser, to our best knowledge.

In this paper, a Tm:YAP laser with the output wavelength of 1989 nm was selected for the first time as the pump source of Q-switched Ho:YAP laser. When the absorbed power was 30 W, the output power of acousto-optic (AO) Q-switched Ho:YAP laser was 18.02 W, and the pulse width was 104.2 ns at repetition frequency of 10 kHz. The corresponding slope efficiency was 70.11%, and the optical-optical conversion efficiency was 43.03%. The output center wavelength was 2129.22 nm.

2. Materials and Methods

The experimental configuration was shown in Figure 1.

To achieve high output power of Q-switched Ho:YAP laser, four semiconductor lasers (Type: SHCC-FCP-60-200-795-S, Shanghai Chuchuang Optical Machinery Technology Co., Ltd., Shanghai, China) with a central wavelength of 795 nm were used as pumping sources of Tm:YAP laser.

Two Tm:YAP crystals with the same parameter were used in the experiment. The Tm:YAP crystal had a cross-section size of 4 mm × 4 mm, a length of 12 mm, and Tm^{3+} doping concentration of 3 at.%. Both ends of the crystal were coated with high transmissivity at 1989 nm and 795 nm. The crystal was wrapped in thick indium foil with the thickness of 0.1 mm and placed in a copper heat sink. The heat sink was cooled by water, which was kept at 18 °C.

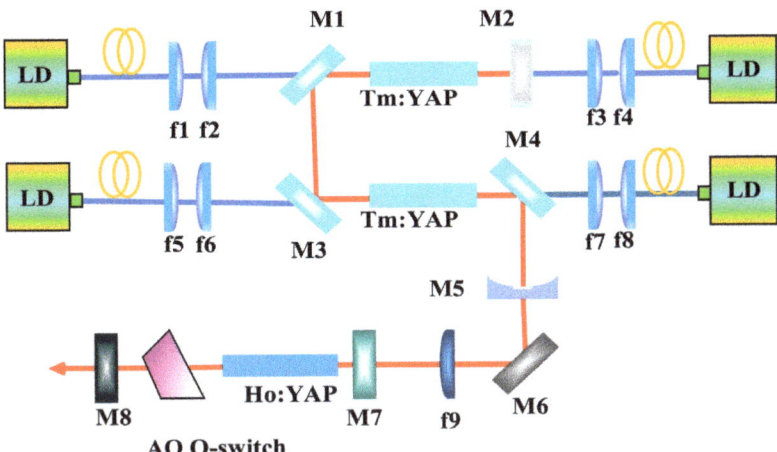

Figure 1. Acousto-optic Q-switched Ho:YAP laser pumped by 1989 nm laser.

The output power of Tm:YAP laser was improved by using a laser diode (LD) double-ended pump structure. The cavity was formed by flat mirrors M1, M2, M3, M4, and a concave mirror M5. M1, M3 and M4 were 45° mirrors coated with high transmissivity at 795 nm and high reflectivity at 1989 nm. M2 was a 0° mirror coated with high transmissivity at 795 nm and high reflection at 1989 nm. Curvature radius of the output coupler M5 was 300 mm and coated with transmissivity of 10% at 1989 nm.

Good mode matching between the pumping beam and oscillating beam of the Tm:YAP laser was achieved by adjusting the focus coupling mirrors, f1 = f4 = f5 = f8 = 25 mm, f2 = f3 = f6 = f7 = 50 mm. The focus lenses were anti-reflection coated at 795 nm. We measured the pump power before the lenses and after M1, M2, M3 and M4, and we calculated the pump transmission to be about 90%.

The resonator of Ho:YAP was a straight cavity composed of M7 and M8. M7 was coated with high transmissivity at 1989 nm and high reflectivity at 2118 nm. Curvature radius of the output coupler M8 was 100 mm and coated with high transmissivity at 1989 nm and transmissivity of 20% at 2118 nm. The cavity length of Ho:YAP was 70 mm.

The size of Ho:YAP crystal was 4 mm × 4 mm × 25 mm, and the Ho^{3+} doped concentration was 0.8 at. %. Both ends of the crystal were coated with high transmissivity at 1989 nm and 2118 nm. The crystal was wrapped in thick indium foil with the thickness of 0.1 mm and placed in a copper heat sink. The heat sink was cooled by water, which was kept at 18 °C.

A quartz acousto-optic Q-switch (QS041-10M-HI8 and the drive model MQH041-100DM-A05, Gooch&Housego Co., Ltd, Ilminster, Somerset, UK) with a length of 46 mm and aperture of 2.0 mm was employed for Q-switching operation. Both ends of the Q-switch crystal were coated with high transmissivity at 2118 nm. The radio frequency was 40.68 MHz, and the maximum radio frequency power was 50 W. The threshold of damage was larger than 500 MW/cm^2. The AO Q-switch crystal was cooled by a water cooler at 18 °C.

In order to facilitate the adjustment and realize the good mode matching between the pump light and the oscillating light, plat mirror M6 and focus lenses f9 and f10 were used. M6 was a 45° full mirror coated with high-reflection at 1989 nm. The focus lens f9 = 50 mm was anti-reflection (AR) coated at 1989 nm.

3. Results and Discussion

The absorptance of Tm:YAP crystal to pump light was 92%. The output power of Tm:YAP laser was measured with the power meter F150A (OPHIR, Jerusalem, Israel), as shown in Figure 2. The output power of the laser varied linearly with the absorbed power. The threshold power of the laser was 11 W. The maximum output power of the Tm:YAP laser was 50 W, and the slope efficiency was 41.32%. The central wavelength at the maximum output power was 1989.01 nm, which was measured using the spectrometer (AQ6370 of Yokogawa, Musashino, Tokyo, Japan), as shown in Figure 3.

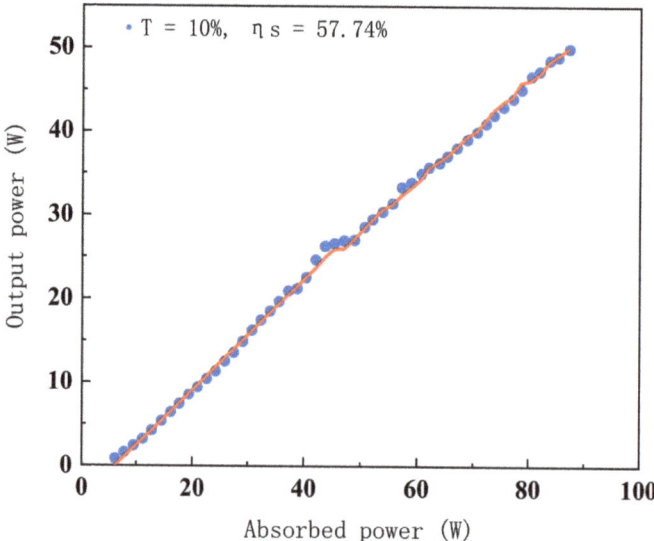

Figure 2. Output power versus absorbed power of Tm:YAP laser.

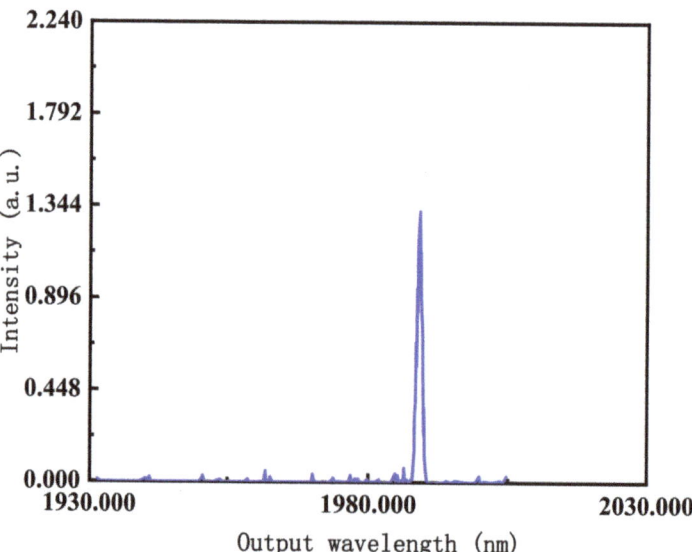

Figure 3. Output spectrum of Tm:YAP laser.

The central wavelengths of Tm:YAP laser versus output power were shown in Figure 4. When the output power of continuous Tm:YAP laser varied from 0.5 W to 50 W, the center wavelength of the Tm:YAP laser remained between 1986.00 nm and 1990.00 nm. The fluctuation of the center wavelength was affected by the accuracy of temperature control of the Tm:YAP crystal. However, the output wavelength of the Tm:YAP laser was always in the absorption line width of the Ho:YAP crystal, which means that it can be used as the pump source of the Ho:YAP laser.

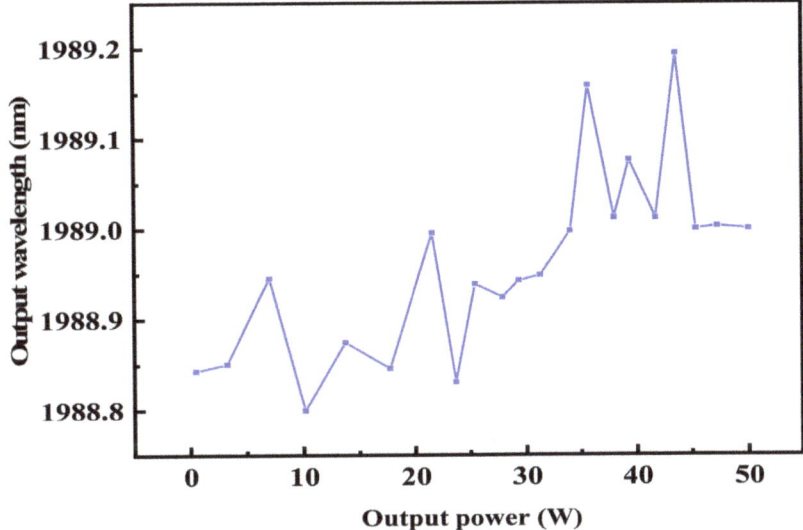

Figure 4. Output wavelength of Tm:YAP laser versus output power.

With Tm:YAP laser as the pump source with the maximum output power of 50 W, the experimental study of acousto-optic Q-switched Ho:YAP laser was carried out. The average power of the laser output was measured by a power meter (30A-BB-18, OPHIR, Jerusalem, Israel), and the pulse width of the laser output was measured by an oscilloscope (DPO3054, Tektronix, Beaverton, Oregon, U.S.) and a pulse width detector (PCI-3TE-12, VIGO System S.A., Warsaw, Poland).

As shown in Figure 5, the average output powers of an AO Q-switched Ho:YAP laser versus pump power were achieved under repetition frequency of 1 kHz, 5 kHz and 10 kHz. At pump power of 50 W, the maximum average output powers of AO Q-switched Ho:YAP laser were 14.2, 15.84, and 18.02 W, with the slope efficiencies of 55.25, 61.66, and 70.11%, respectively. The output pulse width of the Q-switched Ho:YAP laser versus the pump power was achieved under different repetition frequencies, as shown in Figure 6. At absorbed power of 30 W, the narrowest output pulse widths were 101.7 ns, 103.1 ns and 104.2 ns under repetition frequency of 1 kHz, 5 kHz, and 10 kHz, respectively.

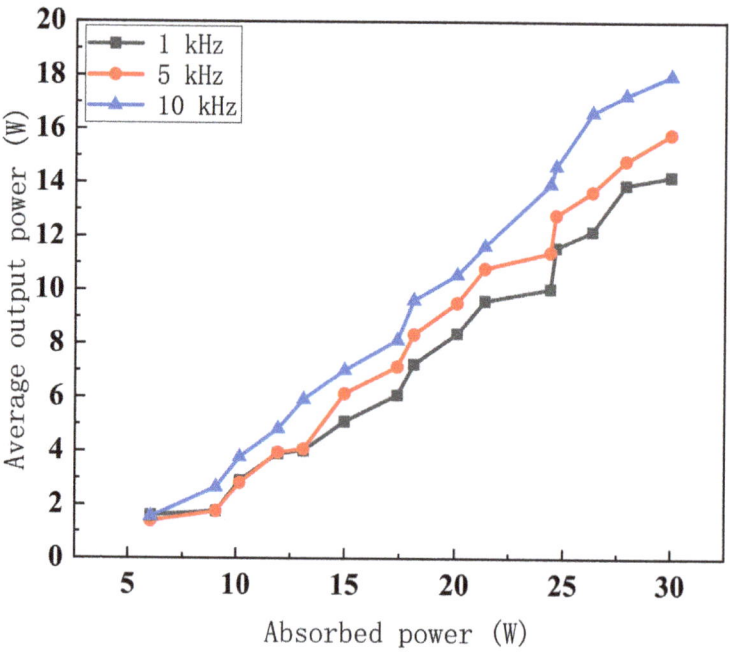

Figure 5. Average output power of Q-switched Ho:YAP laser versus absorbed power.

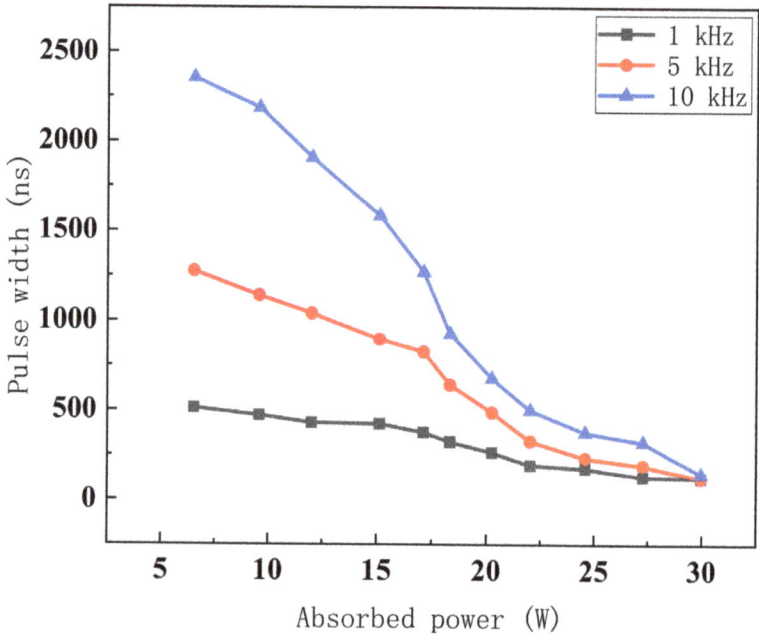

Figure 6. Output pulse width of Q-switched Ho:YAP laser versus absorbed power.

The central wavelength of the AO Q-switched Ho:YAP laser was measured using the spectrometer (AQ6370, Yokogawa, Musashino, Tokyo, Japan). The central wavelengths at the maximum output average power were 2129.29 nm, 2129.47 nm and 2129.22 nm, with output linewidths of 0.77 nm, 0.75 nm and 0.74 nm under repetition frequency of 1 kHz, 5 kHz and 10 kHz, respectively.

Figure 7 showed the output spectrum of a Q-switched Ho:YAP laser at an output average power of 18.02 W and repetition frequency of 10 kHz. While, Figure 8 gave the output width of Q-switched Ho:YAP laser under the same condition.

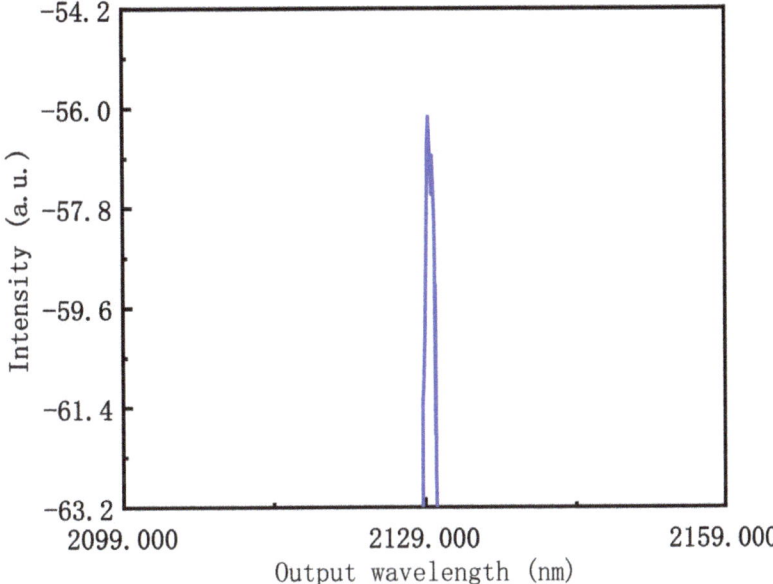

Figure 7. Output spectrum of AO Q-switched Ho:YAP laser.

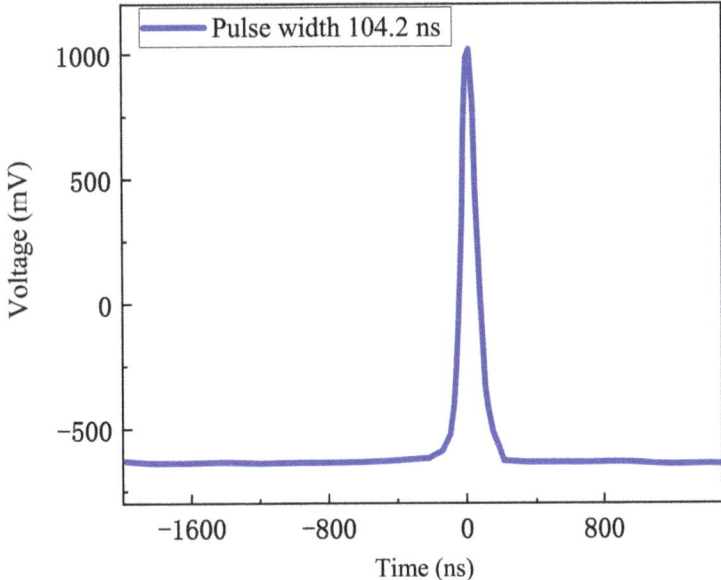

Figure 8. Output pulse width of AO Q-Switched Ho:YAP laser.

4. Conclusions

We demonstrated an AO Q-switched Ho:YAP laser pumped by the 1989 nm laser for the first time. Under the pump power of 50 W, at a PRF of 1 kHz, the average output power of 14.2 W Ho:YAP laser was obtained, with the slope efficiency of 55.25%, the pulse width of 101.7 ns, and the central wavelength of 2129.29 nm. At a PRF of 5 kHz, the average output power of 15.84 W laser was obtained, with the slope efficiency of 61.66%, the pulse width of 103.1 ns, and the central wavelength of 2129.47 nm. At a PRF of 10 kHz, the average output power of 18.02 W laser was obtained, with the slope efficiency of 70.11%, the pulse width of 104.2 ns, and the central wavelength of 2129.22 nm.

Author Contributions: Conceptualization, C.N., Y.W. and C.W.; methodology, T.D. and C.N.; software, C.N., Y.W. and L.Z.; validation, C.N. and Y.W.; formal analysis, X.C. and L.Z.; investigation, Y.J., C.N., L.Z. and X.C.; resources, C.W. and T.D.; data curation, Y.W., Y.J. and C.N.; writing—original draft preparation, C.N.; writing—review and editing, C.W. and Y.W.; visualization, C.N.; supervision, T.D.; project administration, C.W.; funding acquisition, C.W. All authors have read and agreed to the published version of the manuscript.

Funding: This research was funded by Study on the radiation mechanism and output characteristics of single longitudinal mode laser with tunable injection frequency locked 2 μm pulses, grant number 202002041JC and the APC was funded by Science and Technology Department of Jilin Province in China.

Data Availability Statement: The data presented in this study are available on request from the corresponding author.

Acknowledgments: This work is supported by Science and Technology Department of Jilin Province in China (Grant No. 202002041JC).

Conflicts of Interest: The authors declare no conflict of interest.

References

1. Wu, C.T.; Jiang, Y.; Dai, T.Y.; Zhang, W.Q. Research Progress of 2 μm Ho-doped Solid-state Laser. *Chin. J. Lumin.* **2018**, *39*, 1584–1597.
2. Scholle, K.; Lamrini, S.; Koopmann, P. *2 μm Laser Sources and Their Possible Applications In Frontiers in Guided Wave Optics and Optoelectronics*; IntechOpen: London, UK, 2010; Volume 21, pp. 471–500.
3. Li, Z.; Heidt, A.M.; Daniel, J.M.O.; Jung, Y.; Alam, S.U.; Richardson, D.J. Thulium-Doped Fiber Amplifier for Optical Communications at 2 Microns. *Opt. Express* **2013**, *21*, 9289–9297. [CrossRef] [PubMed]
4. Lombard, L.; Valla, M.; Augère, B.; Planchat, C.; Goular, D.; Bourdon, P.; Canat, G. Eyesafe Coherent Detection Wind Lidar Based on a Beam-Combined Pulsed Laser Source. *Opt. Lett.* **2015**, *40*, 1030–1033. [CrossRef] [PubMed]
5. Gibert, F.; Edouart, D.; Cénac, C.; Le Mounier, F. 2-μm High-Power Multiple-Frequency Single-Mode Q-Switched Ho:YLF Laser for DIAL Application. *Appl. Phys. B* **2014**, *116*, 967–976. [CrossRef]
6. Peplow, P.V.; Chung, T.Y.; Baxter, G.D. Laser Photostimulation (660 nm) of Wound Healing in Diabetic Mice Is Not Brought About by Ameliorating Diabetes. *Lasers Surg. Med.* **2012**, *44*, 26–29. [CrossRef] [PubMed]
7. Wu, Y.; Zhai, G.; Yao, Z.H. Development of 2μm Band Lasers. *Laser J.* **2008**, *29*, 3–4.
8. Ding, Y. Characteristics of 2 μm Single Doped Ho Vanadate Solid State Laser. Ph.D. Thesis, Harbin Institute of Technology, Harbin, China, 2015.
9. Xin, Y.; Ye, B.; Fang, W.L. Application and progress of holmium laser. *Laser Optoelectron. Prog.* **2012**, *49*, 22–27.
10. Yang, X.T. Room Temperature Resonant Pumping Ho: Experimental Study of YAP Laser. Ph.D. Thesis, Harbin Institute of Technology, Harbin, China, 2009.
11. Yu, T.; Peng, Y.J.; Ye, X.S.; Chen, W.B. Study on Ho: YAP laser technology pumped by thulium optical laser. In *Seminar on Optical Technology 2017 and Collection of Interdisciplinary Forum*; Shanghai Infrared and Remote Sensing Society: Shanghai, China; Yunnan Optical Society: Yunnan, China, 2017; Volume 6.
12. Wu, X.S. Research on Ho: YAP Single-Frequency Laser. Ph.D. Thesis, Harbin Institute of Technology, Harbin, China, 2019.
13. Duan, X.M.; Yao, B.Q.; Li, G.; Wang, T.H.; Yang, X.T.; Wang, Y.Z.; Zhao, G.J.; Dong, Q. High Efficient Continuous Wave Operation of a Ho: YAP Laser at Room Temperature. *Laser Phys. Lett.* **2009**, *6*, 279–281. [CrossRef]
14. Yang, X.T.; Ma, X.Z.; Li, W.H.; Liu, Y. Q-Switched Ho: YAlO$_3$ Laser Pumped by Tm:YLF Laser at Room Temperature. *Laser Phys.* **2011**, *21*, 2064–2067. [CrossRef]
15. Yang, X.T.; Liu, Y.; Li, W.H.; Ju, Y.L. Theoretical and Experimental Analysis of 2 μm Laser Crystal Ho:YAP. *Infrared Laser Eng.* **2012**, *41*, 1733–1737.

16. Dai, T.Y.; Ju, Y.L.; Shen, Y.J.; Wang, W.; Yao, B.Q.; Wang, Y.Z. High-Efficiency Continuous-Wave and Q-Switched Operation of a Resonantly Pumped Ho:YAP Ring Laser. *Laser Phys.* **2012**, *22*, 1292–1294. [CrossRef]
17. Wang, Z.; Ma, X.; Li, W. Efficient Ho: YAP Laser Dual-End-Pumped by Tm Fiber Laser. *Opt. Rev.* **2014**, *21*, 150–152. [CrossRef]
18. Duan, X.M.; Yang, C.H.; Shen, Y.J.; Yao, B.Q.; Ju, Y.L.; Wang, Y.Z. High-Power in-Band Pumped a -Cut Ho: Yap Laser. *J. Russ. Laser Res.* **2014**, *35*, 239–243. [CrossRef]
19. Duan, X.M.; Lin, W.M.; Cui, Z.; Yao, B.Q.; Li, H.; Dai, T.Y. Resonantly Pumped Continuous-Wave Mode-Locked Ho:YAP Laser. *Appl. Phys. B* **2016**, *122*, 88. [CrossRef]
20. Duan, X.; Li, L.; Shen, Y.; Yao, B. Efficient Ho:YAP Laser Dual End-Pumped by a Laser Diode at 1.91 μm in a Wing-Pumping Scheme. *Appl. Phys. B* **2018**, *124*, 1–6. [CrossRef]
21. Guo, L.; Zhao, S.; Li, T.; Qiao, W.; Ma, B.; Yang, Y.; Yang, K.; Nie, H.; Zhang, B.; Wang, R.; et al. In-band Pumped, High-Efficiency LGS Electro-Optically Q-Switched 2118 nm Ho:YAP Laser with Low Driving Voltage. *Opt. Laser Technol.* **2020**, *126*, 106015. [CrossRef]

Article

Experimental Study of Plasma Plume Analysis of Long Pulse Laser Irradiates CFRP and GFRP Composite Materials

Yao Ma, Chao Xin, Wei Zhang and Guangyong Jin *

Jilin Key Laboratory of Solid-State Laser Technology and Application, School of Science, Changchun University of Science and Technology, Changchun 130022, China; mayao@cust.edu.cn (Y.M.); xinchao@cust.edu.cn (C.X.); zhangwei@cust.edu.cn (W.Z.)
* Correspondence: jgyciom@cust.edu.cn; Tel.: +86-431-85582465

Abstract: The application of laser fabrication of fiber-reinforced polymer (FRP) has an irreplaceable advantage. However, the effect of the plasma generated in laser fabrication on the damage process is rarely mentioned. In order to further study the law and mechanism of laser processing, the laser process was measured. CFRP and GFRP materials were damaged by a 1064 nm millisecond pulsed laser. Moreover, the propagation velocity and breakdown time of plasma plume were compared. The results show that GFRP is more vulnerable to breakdown than CFRP under the same conditions. In addition, the variation of plasma plume and material surface temperature with the number of pulses was also studied. The results show that the variation trend is correlated, that is, the singularities occur at the second pulse. Based on the analysis of experimental phenomena, this paper provides guidance for plasma phenomena in laser processing of composite materials.

Keywords: long pulse laser; plasma plume; composite materials; CFRP; GFRP

1. Introduction

A composite material is usually defined as a combination of two or more different materials that have different physical and chemical properties. Thus, composite materials may possess the superior properties of both components [1,2]. Generally, in the fiber-reinforced polymer (FRP), fiber wrapped by the resin matrix is the main load-bearing component of the material. At the same time, the resin is the load transfer element, which also plays the role of protecting the whole structure [3,4].

The properties of high specific strength and high modulus make fiber-reinforced resin composites useful in many fields, such as military, aviation, and sport equipment [5]. Because this kind of material is composed of fiber and matrix layering, there are several factors that influence the stiffness and strength of the composite material. For example, the direction of the fiber, the weaving method, the layering method, and the number of layers. Theoretically, there are many reinforcements and matrix materials that can be combined in a huge number of ways to produce composite structures that match specific applications of the structure. Fiber-reinforced lamination is composed of multilayer unidirectional or bidirectional fibers located within the matrix. Normally, materials should have similar properties (isotropy) in different directions, that is, they should have the same properties (homogeneity) at any point in the material. This means that the properties of the material are the same regardless of the direction and position of the material. However, when they are glued together and new special structures emerge, the properties are different [6]. Unidirectional fiber-reinforced materials exhibit different properties in the transverse direction of the fibers, while bidirectional materials exhibit different properties in all directions [7–12].

In the traditional middle infrared thermal imaging technology, the thermal imager records the mid-infrared radiation emitted by the object itself, which contains the characteristic information of the material; the difference of the mid-infrared radiation ability of material surfaces is caused by the change of the temperature field of the material surface. Usually, in the absence of external thermal excitation, the internal loss of the material will be reduced. Defects could not cause the temperature field change of the material surface, so the passive thermal imaging technology could not detect the defects under the material surface. Different from the traditional passive mid-infrared thermal imaging technology, the thermal wave detection technology uses the active control of light, heat, and other ways to stimulate the defects. The results show that the physical properties of different material surfaces and under the surface will affect the transmission of the heat wave and reflect the change of temperature field on the material surface in some way. By controlling the thermal excitation method and recording the infrared radiation of the material surface, the nonuniformity of the material can be obtained [13,14].

In order to achieve the purpose of detection and flaw detection, the mechanism of laser induced damage of CFPR needs to be further explored. These processes of laser irradiation of fiber-reinforced polymers is very complicated, which is different from that of isotropic material or unistructural material. First, the material is heated by a laser, and the surface of the composite material is heated rapidly. Then, the surface of the resin reaches its decomposition temperature, its chemical bonds break, and decomposition occurs. Then, the material continues to absorb laser energy [15], resulting in combustion or gasification. Finally, the laser disintegrates and ionizes the combustion products to form plasma.

Therefore, in this paper, the experimental content proposed is to measure the temperature and combustion wave changes of FRP materials under laser damage in real time to obtain the process and law of long pulse lasers acting on FRP materials. In addition, the effects of different components, different structures, and different laser parameters on processing are compared, so as to provide technical reference for long pulse laser processing of FRP materials.

2. Materials and Methods

Melar 10, Nd: YAG, a millisecond pulse laser (Beamtech Optronics Co., Ltd, Beijing, China), was set up for the emitting laser. The laser wavelength was 1064 nm, the pulse frequency was 10 Hz, the pulse width was 0.5–3 ms, and the energy was variable from 0.5–10 J. The laser was partially reflected to the energy meter probe after passing the beam splitting mirror to measure the laser energy of the material in real time. At the same time, a KMGA740 high-speed thermometer was used to measure the temperature at the center point of the material surface. Meanwhile, the plasma generated in the experiment was photographed using the shadow method using the Phamton high speed camera (Vision Research, Inc., New York, NY, USA). The reference light was the 532 nm green laser, which propagated from the beam expanding mirror and the focusing lens. The frame frequency of the high-speed camera was 18,000 fps, and the exposure time was 10 s (see Figure 1).

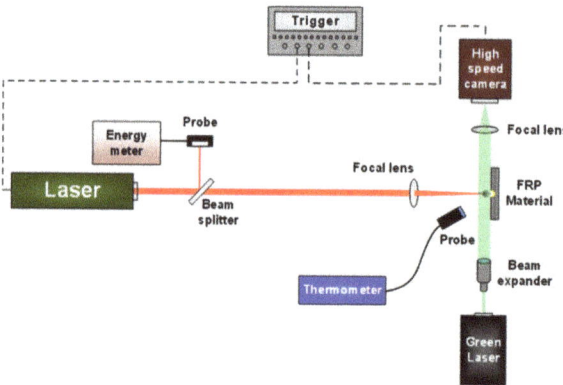

Figure 1. Experimental set-up.

Carbon fiber-reinforced polymer (CFRP) and glass fiber-reinforced polymer (GFRP) were used in the experiment. The gasification temperatures of the two materials are 3500 °C and 1000 °C, respectively. At the same time, the material structure is a multi-layer orthogonal bidirectional and unidirectional structure. The material is black sheet, which is composed of a thin layer of fiber and epoxy resin superimposed and solidified, with a surface size of 50 mm × 50 mm and a thickness of 2 mm.

3. Results and Discussion

3.1. Plasma Plume Analysis

When multiple laser pulses were applied to the composite surface, the material was broken down. A high-speed camera was used to take pictures of the laser action process to obtain the corresponding pulse number of the material breakdown (the frame frequency of the high-speed camera was 18,000 fps, and the exposure time was 10 s) and to analyze the combustion wave morphology. At the same time, the propagation velocity of combustion waves on the front and back surfaces of the materials was measured to obtain the propagation law of combustion waves.

The shadow of ejected vapor does not appear on the back surface of the material, as shown in Figure 2a. However, As can be seen from the morphology, tense ejected material appears on both CFRP and GFRP surfaces, and this part of the material not only exists on the front surface but also appears on the back surface of the material, as shown in Figure 2b. This is because when the 14th pulse acted on CFRP, the plasma and spatter concentration on the back surface was low, which had dissipated after the end of the laser pulse, so no significant shadow was formed.

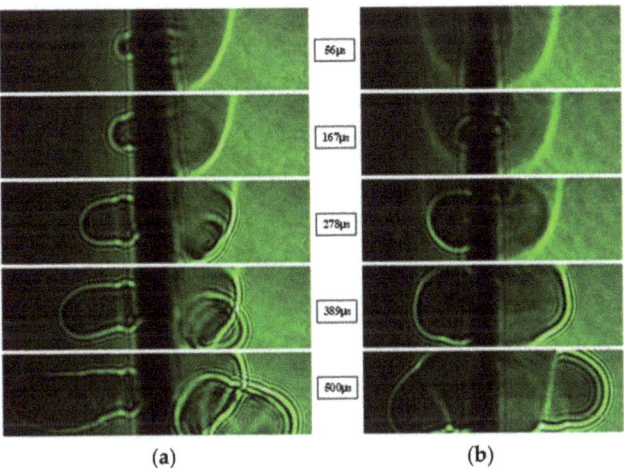

Figure 2. Combustion wave morphology (a) unidirectional CFRP, $\tau p = 1$ ms, $N = 20$, $n = 14$ th, $d = 2$ mm, $F = 392$ J/cm^2. (b) Unidirectional GFRP, $\tau p = 1$ ms, $N = 20$, $n = 13$th, $d = 2$ mm, $F = 392$ J/cm^2.

At the same time, it can be seen in Figure 2a that when the laser irradiated on the material surface with ejected vapor, the plasma expands slowly. When the combustion wave keeps moving outside the ejected vapor, its motion direction changes. This is because the interior structure of the ejected material is complex and uneven. When the laser penetrates the surface of the material, it is no longer the original waveform. At the same time, the process of outward expansion is also affected by the internal components of the ejected material, so the angle is offset.

Figure 3a,b show the combustion wave expansion velocity of CFRP materials during 20 pulses. Under the same laser conditions, the bidirectional material is penetrated at the 12th pulse and unidirectional at the 10th pulse. In the process of laser irradiating, the expansion velocity on the back surface is slow in the initial period and increases in the later period. In the early period of material penetration, only few laser energies can penetrate the plasma inside the keyhole and act on the back surface, thus producing insufficient energy. When the backside material is badly ablated, more laser energy can penetrate the material and act on the plasma on the back surface, so the combustion wave velocity is faster. This process is similar to the situation on the front surface in which the speed first increases and then decreases. This phenomenon conforms to the law of energy conservation. At the same time, by comparing the two different structures, the material of the bidirectional structure was broken down in the 12th pulse, while the unidirectional material was broken down in the 10th pulse. From Figure 3c,d, due to the number of pulses, the bidirectional structure is less likely to be damaged for the CFRP material.

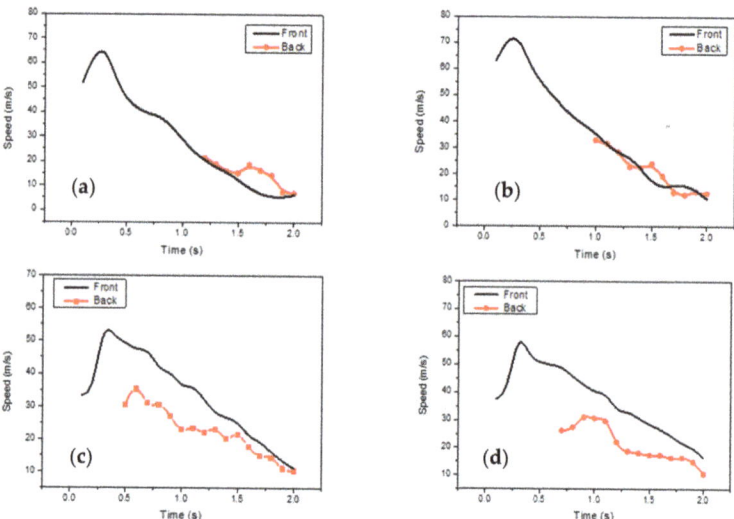

Figure 3. Combustion wave expansion velocity changing with time (τp = 1 ms, N = 20, d = 2 mm, F = 392 J/cm^2). (**a**) Bidirectional CFRP; (**b**) Unidirectional CFRP; (**c**) Bidirectional GFRP; (**d**) Unidirectional GFRP.

Figure 3c,d show the combustion wave expansion velocity of GFRP materials during 20 pulses. By comparing that to the CFRP materials, the trend is similar. However, bidirectional material is broken at the 5th pulse, while the unidirectional GFRP is broken at the 7th pulse. Under the same laser conditions, the bidirectional structure GFRP material is more easily broken down.

Figure 4 shows the relationship of combustion wave expansion velocity changing with time of FRP materials, and the dots in the figure show the breakdown time of the material. By comparing the action law before breakdown, the plasma expansion velocity of CFRP material decreases with the increase in the number of pulses and only increases when the second or third pulse occurs. However, GFRP material increases with the number of pulses. By comparing the breakdown time of the two materials, GFRP material was broken before CFRP material, regardless of the material structure. In addition, the comparison of the expansion velocity shows that the velocity of CFRP in the early stage is greater than that of GFRP, and that of GFRP in the late stage is greater than that of CFRP.

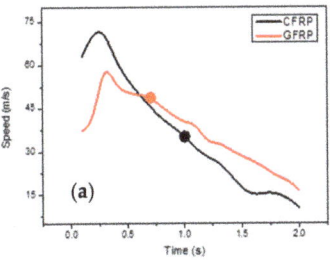 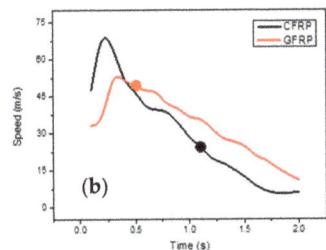

Figure 4. Combustion wave expansion velocity changing with time (τp = 1 ms, N = 20, d = 2 mm, F = 392 J/cm^2), (**a**) Unidirectional; (**b**) Bidirectional.

3.2. Temperature Analysis

In the experiment, the surface area of material was measured. Under the action of the multi-pulse laser, the material is injected with laser energy, so that the temperature of the material rises until decomposition, gasification, and even ionization. When the pulse laser fluence is 392 J/cm^2, a single pulse can gasify and ionize the FRP material, resulting in an ablative zone. Therefore, plasma is sensitive to laser conditions; because of the generation time of the plasma difference in the two materials the plasma is sometimes beneficial to heat conduction, but sometimes it can be the opposite due to the plasma shielding effect.

In laser processing of composite materials, a multi-pulse laser is often used to irradiate them to improve the processing efficiency by utilizing the accumulation effect of temperature and plasma. Therefore, in the research of multi-pulse laser processing, the residual temperature of the material surface after each pulse ends while the next pulse arrives is considered to represent the temperature accumulation effect of the material, so as to analyze the temperature during the processing of the material.

Figure 5 shows the curve of the residual temperature of CFRP and GFRP during 20 laser pulses. The overall trend is that the residual temperature increases with the number of pulses, and the accumulative trend is obvious. However, at the same laser fluence, the residual temperature of GFRP material is higher than that of CFRP, and the temperature of GFRP decreases in the second pulse, while that of CFRP continues to rise.

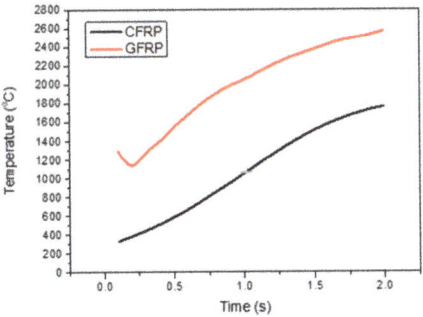

Figure 5. Residual temperature changing with time when laser damaged FRP materials (τp = 1 ms, N = 20, d = 2 mm, F = 392 J/cm^2).

From the figure, the residual temperature of the two materials is quite different. The CFRP materials fall in temperature below the gasification temperature (3500 °C), which is higher than the plasma temperature (104 °C). This shows that when the next pulse arrives, plasma dissipated toward an inverse direction of laser propagation. Thus, the plasma density is low, which cannot form the plasma shielding effect of the laser. On the contrary, the residual temperature of GFRP material is much higher than that of the material gasification point (1000 °C), so the measured temperature should be a mixture of material decomposition and plasma. This is because when compared to carbon fiber, glass fiber is mainly composed of SiO_2, which is usually amorphous and has a higher melting point and damage threshold. Thus, GFRPs are more likely to produce a higher concentration of the mixture than CFRPs, while the mixture shields the incident laser. After that, laser energy irradiated on the composite material surface decreases, and the cumulative effect becomes weak in a short period of time. However, when the subsequent pulsed laser continues to irradiate, the mixture in front of the material surface is ignited by the laser, and the combustion wave propagates in the opposite direction, which weakens the shielding effect. Thus, the laser energy can re-pass through the mixture and act on the surface of the material, thus intensifying the effect and promoting the temperature accumulation.

3.3. Correlation Analysis

Based on the analysis of plasma plume and temperature, these two dependent variables both appear as singularities with the increase in the number of pulses. However, the difference is that for CFRP, the residual temperature curve is relatively smooth and there is no singularity. This is because the damage mechanism of the two materials is different when they are damaged by laser. Comparing the material properties of carbon fiber and glass fiber, we can see that the damage mechanism of carbon fiber and glass fiber is quite different.

The epoxy resin attached to the surface of the fiber has a higher laser transmittance than the fiber, so the laser directly acts on the fiber through the epoxy resin on the surface of the material [8]. Carbon fiber is an opaque medium, while glass fiber is a translucent medium. This difference results in the different transmittance of the material to the laser, that is, the transmittance of glass fiber is much higher than that of carbon fiber.

The electronic drift in CFPR and GFPR is disordered. The main reason for the difference between the two curves is lattice–lattice coupling. It is well known that the lattice of carbon in CFPR is periodic, while that of SiO_2 in GFPR is aperiodic. Therefore, the thermal conductivity of the two materials is different under the same laser conditions.

Comparing the two kinds of fibers, the thermal conductivity of carbon fiber is 60 W/m·K, while that of glass fiber is only 1.09 W/m·K. According to the above two parameters, the speed of CFRP rising to the gasification point is much higher than that of GFRP during the heating process of laser activated material. However, the gasification point of the two is different. As we discussed above, glass fiber is mainly composed of SiO_2, which is usually amorphous, it has no fixed melting point; when the same energy laser acts on the surface of the material and causes the material to heat up, the glass fiber is more likely to heat up to its gasification temperature and ionize. Glass fiber or other forms tend to form stable melts within a large range, and also change their viscosity gradually before they reach a liquid state. Carbon fiber tends to oxidize in air and pyrolysis in Ar atmospheres. In fact, the absorption of pure glass at 1064 nm is relatively low [16,17]. The plasma temperature produced by ionization is much higher than that of the surface temperature of the material. Meanwhile, based on the thermal conductivity factor, the cooling rate of CFRP materials is also faster. Therefore, at the end of each pulse, the residual temperature of CFRP is much lower than GFRP.

The plasma and temperature changes of the two materials were compared. There are singularity effects on temperature and plasma plume of GFRP materials. At the second pulse, the plume has a higher velocity and a lower temperature. Both defy the overall trend. When the laser acts on the material, the fiber acts as the main medium for heat exchange. Its temperature is first raised, and then the heat is transferred to the surrounding epoxy resin through heat conduction. Therefore, at the first pulse, although the temperature of the fiber first rises, it does not reach the gasification point, but the surrounding epoxy resin has decomposed and ionized. Under the continuous action of the laser, the fiber is vaporized and decomposed along with the epoxy, resulting in a flocculent mixture that floats on the surface of the material, which is dense and difficult to disperse. When the second pulse acts on the material, the mixture on the surface first receives the laser energy for secondary decomposition and ionization, which accelerates the propagation speed of the plasma plume. At the same time, due to the shielding of this part of the mixture, the material table can receive less laser energy, so there is an obvious cooling phenomenon. After that, when the subsequent pulses arrive, the shield on the surface of the material is thin because the previously thick mixture has largely decomposed and ionized. While there is still a case for absorbing subsequent lasers, most laser pulses can penetrate the shield and hit the material. Therefore, when the laser distribution of the two parts reaches a relatively balanced state, the subsequent variation trend of plasma plume and temperature tends to be monotonous.

4. Conclusions

In this paper, the temperature and combustion wave changes in FRP materials processed by a long pulse infrared laser are studied.

Under the multi-pulse conditions, the residual temperature of the material increased with the increase in the number of pulses for the two materials. In addition, the morphology of the plasma combustion wave was also analyzed when the material was broken down, and the phenomenon of the ejected vapor shadow and the deviation of the combustion wave propagation direction was found. However, the residual temperature of GFRP material was higher than that of CFRP under the same laser fluence. Moreover, the residual temperature of CFRP continued to rise, but the temperature of GFRP dropped in the second pulse due to the shielding effect. Finally, under the same laser conditions, CFRP material of a bidirectional structure was less likely to be damaged by the pulse train laser, while GFRP material of a bidirectional structure was more likely to be broken down. Due to the different structure of the materials, when the multi-pulse laser irradiates the surface of the carbon fiber material, the glass fiber material was decomposed before the carbon fiber material. As a result, when the laser parameters are precisely controlled, multi-pulse laser can process FRP materials more effectively due to the temperature accumulation effect and the formation of plasma.

Author Contributions: Conceptualization, Y.M. and C.X.; methodology, W.Z.; software, W.Z.; validation, Y.M., C.X., and W.Z.; formal analysis, Y.M.; investigation, Y.M.; resources, Y.M.; data curation, Y.M.; writing—original draft preparation, G.J.; writing—review and editing, Y.M.; visualization, Y.M.; supervision, Y.M.; project ad funding acquisition, G.J. All authors have read and agreed to the published version of the manuscript.

Funding: This research was funded by National Natural Science Foundation of China, grant number U19A2077 and the Science and Technology Research project of the Jilin Provincial Department of Education (Grant No. JJKH20200732KJ).

Data Availability Statement: The data presented in this study are available on request from the corresponding author.

Acknowledgments: We thank the Jilin Key Laboratory of Solid-State Laser Technology and Application, School of Science, Changchun University of Science and Technology.

Conflicts of Interest: The authors declare no conflict of interest.

References

1. Rahman, M.; Ramakrishna, S.; Prakash, J.R.S.; Tan, D. Machinability study of carbon fiber reinforced composite. *J. Mater. Process. Technol.* **1999**, *89*, 292–297. [CrossRef]
2. William, F.; Smith Hashemi, J.; Presuel-Moreno, F. *Foundations of Materials Science and Engineering*; Mcgraw-Hill Publishing: New York, NY, USA, 2006.
3. King, R.L. Fibre-Reinforced Composites Materials, Manufacturing and Design. *Composites* **1989**, *20*, 172–173. [CrossRef]
4. Mitchell, B.S. *An Introduction to Materials Engineering and Science for Chemical and Materials Engineers*; John Wiley & Sons: Hoboken, NY, USA, 2004.
5. Karataş, M.A.; Gökkaya, H. A review on machinability of carbon fiber reinforced polymer (CFRP) and glass fiber reinforced polymer (GFRP) composite materials. *Def. Technol.* **2018**, *14*, 318–326. [CrossRef]
6. Yung, K.C.; Mei, S.M.; Yue, T.M. A study of the heat-affected zone in the UV YAG laser drilling of GFRP materials. *J. Mater. Process. Technol.* **2002**, *122*, 278–285. [CrossRef]

7. Ohkubo, T.; Sato, Y.; Matsunaga, E.; Tsukamoto, M. Three-dimensional numerical simulation during laser processing of CFRP. *Appl. Surf. Sci.* **2017**, *417*, 104–107. [CrossRef]
8. Takahashi, K.; Tsukamoto, M.; Masuno, S.; Sato, Y. Heat conduction analysis of laser CFRP processing with IR and UV laser light. *Compos. Part A Appl. Sci. Manuf.* **2016**, *84*, 114–122. [CrossRef]
9. Abrao, A.M.; Rubio, J.C.; Faria, P.E.; Davim, J.P. The effect of cutting tool geometry on thrust force and delamination when drilling glass fibre reinforced plastic composite. *Mater. Des.* **2008**, *29*, 508–513. [CrossRef]
10. Tsao, C.C.; Hocheng, H. Effect of tool wear on delamination in drilling composite materials. *Int. J. Mech. Sci.* **2007**, *49*, 983–988. [CrossRef]
11. Durão, L.M.P.; Gonçalves, D.J.; Tavares, J.M.R.; de Albuquerque, V.H.C.; Vieira, A.A.; Marques, A.T. Drilling tool geometry evaluation for reinforced composite laminates. *Compos. Struct.* **2010**, *92*, 1545–1550. [CrossRef]
12. Goeke, A.; Emmelmann, C. Influence of laser cutting parameters on CFRP part quality. *Phys. Procedia* **2010**, *5*, 253–258. [CrossRef]
13. Abrão, A.M.; Faria, P.E.; Rubio, J.C.; Reis, P.; Davim, J.P. Drilling of fiber reinforced plastics: A review. *J. Mater. Process. Technol.* **2007**, *186*, 1–7. [CrossRef]
14. Zhukova, E.S.; Zhang, H.; Martovitskiy, V.P.; Selivanov, Y.G.; Gorshunov, B.P.; Dressel, M. Infrared Optical Conductivity of Bulk Bi$_2$Te$_2$Se. *Crystals* **2020**, *10*, 553. [CrossRef]
15. Liu, J.-X.; Mei, S.-L.; Chen, X.-H.; Yao, C.-J. Recent Advances of Near-Infrared (NIR) Emissive Metal Complexes Bridged by Ligands with N- and/or O-Donor Sites. *Crystals* **2021**, *11*, 155. [CrossRef]
16. Huang, Z.Q.; Hong, M.H.; Do, T.B.M.; Lin, Q.Y. Laser etching of glass substrates by 1064 nm laser irradiation. *Appl. Phys. A* **2008**, *93*, 159–163. [CrossRef]
17. Xu, H.B.; Hu, J.; Yu, Z. Absorption behavior analysis of Carbon Fiber Reinforced Polymer in laser processing. *Opt. Mater. Express.* **2015**, *5*, 2330–2336. [CrossRef]

MDPI
St. Alban-Anlage 66
4052 Basel
Switzerland
Tel. +41 61 683 77 34
Fax +41 61 302 89 18
www.mdpi.com

Crystals Editorial Office
E-mail: crystals@mdpi.com
www.mdpi.com/journal/crystals